Robert Lambert Playfair

The Fishes of Zanzibar

Acanthopterygii

Robert Lambert Playfair

The Fishes of Zanzibar
Acanthopterygii

ISBN/EAN: 9783744762939

Printed in Europe, USA, Canada, Australia, Japan

Cover: Foto ©berggeist007 / pixelio.de

More available books at **www.hansebooks.com**

THE

FISHES OF ZANZIBAR

ACANTHOPTERYGII

BY

LIEUT. COLONEL R. LAMBERT PLAYFAIR,
HER MAJESTY'S POLITICAL AGENT AND CONSUL AT ZANZIBAR;

PHARYNGOGNATHI ETC.

BY

ALBERT C. L. G. GÜNTHER, M.A., Ph.D., M.D.

LONDON:
JOHN VAN VOORST, PATERNOSTER ROW.
MDCCCLXVI.

ALERE FLAMMAM.

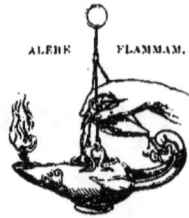

PRINTED BY TAYLOR AND FRANCIS,
RED LION COURT, FLEET STREET.

PREFACE.

A FEW words of explanation as to our object in publishing an account of the Fishes of Zanzibar appear necessary. The east coast of Africa may, for purposes of ichthyology, be conveniently divided into four regions. The first is contained within the limits of the Red Sea; the second extends thence to the Rovuma River, the southern boundary of Zanzibar; the third includes the Portuguese province of Mozambique; and the fourth the British settlements of Natal and the Cape.

The Fish-fauna of three of these has been more or less completely worked out, viz. the Red Sea by Forskål and Rüppell, Mozambique by Peters, and the south-eastern parts by Sir Andrew Smith; while the ichthyology of the great islands lying off the coast has formed the subject of papers by Liénard, Bennett, Guichenot, and others, not to mention the numerous species described by Lacépède, Cuvier, and Valenciennes.

But no attempt has been made to illustrate the Fish-fauna of that large extent of coast stretching between the Straits of Bab-el-Mandeb and Mozambique. The labours of Lieut. Colonel Playfair go far to supply this hiatus. In the course of a residence of many years at Aden and Zanzibar, during which he made frequent excursions to the African coast and the adjacent islands, he formed a considerable collection of Fish, of which the following pages contain a description. With the exception of one or two species described from specimens in the Vienna Museum, which we have not seen, and a few collected by Dr. Kirk in the Rovuma River, now in the British Museum, we have limited ourselves strictly to that collection. We have added, however, a

A 2

nominal list of all the species recorded as having been observed on the East African coast, between the Red Sea on the north and the colony of Natal on the south, including those from Madagascar, Mauritius, and the other islands belonging geographically to the African continent.

The Zanzibar dominions comprise that portion of the coast included between Magdashoa in 2° north latitude, and Cape Delgado in 10° 42' south latitude. Beyond them, to the north, are the independent Somalee tribes, which extend almost to the Red Sea, where they meet the Dankalic race; and on the south they are bounded by Mozambique. The extent of coast under the dominion of the Sultan of Zanzibar is about eleven hundred miles; but the most valuable parts of his sultanate are the islands of Zanzibar (containing the capital of the same name), Pemba, and Monfia. The first is situated at a distance of from twenty to thirty miles from the mainland, and is in size about equal to the Isle of Wight. It contains none but small streams, and, as may be expected, the number of freshwater fishes is very limited. Indeed, excluding those which frequent both salt and fresh water, such as *Gobius* and *Eleotris*, there are only two species of freshwater fish found in the island of Zanzibar, *Clarias gariepinus* and *Fundulus orthonotus*. The large rivers on the mainland are far richer in species; but they have been very imperfectly explored, and still offer a most interesting field for scientific research.

This collection contains 500 distinct species, obtained in the following localities:—

At Zanzibar	428
Mozambique	10
Seychelles	27
Comoro Islands	11
Aden and its vicinity	23
Chagos archipelago	1

Of the 428 observed at Zanzibar, 192 have been noticed in and near the Red Sea, 300 in the Indian Ocean and the seas of the more remote East, 108 on the shores of the various islands lying off the African coast, 7 in the Cape Seas, 3 in the Mediterranean, 25 in the Atlantic Ocean, while 63 appear not to have been found in other localities than the African coast or the rivers which there debouch into the sea.

We have thought it advisable that there should be only one authority for new species, on which account each of the authors has attached his name to a moiety of the work; but it must not be imagined that we have worked otherwise than jointly and continuously throughout.

The numbers in brackets which follow the specific names are those originally attached to the specimens in Lieut. Colonel Playfair's collection. We have retained them to facilitate the identification of such as have been sent by him to other Museums in Europe and India.

We desire to record our grateful thanks to the Government of Bombay for the liberal measure of assistance it has accorded to this work by taking 100 copies. This has enabled us to illustrate all the most important species, and for that end to secure the valuable services of Mr. Ford, whose ability as a zoological artist is too well known to require notice.

THE AUTHORS.

London, November 1866.

LIST OF FISHES

OBSERVED ON THE

EAST COAST OF AFRICA.

ERRATA.

Page 19, line 23, *for* two-fifths *read* very small.
„ 19, line 24, *for* dorsal *read* ventral.
„ 26, line 21, *for* forehead *read* occiput.
„ 26, line 24, *for* anal *read* dorsal.
.. 49, *before* SYNANCEIA *insert* Family COTTIDÆ; *and before* DACTYLOPTERUS *insert* Family CATAPHRACTI. Both these genera should follow PE-DICULATI, p. 70.

THE FISHES OF ZANZIBAR.

Order ACANTHOPTERYGII.

Family PERCIDÆ.

APRION, *Cuv. & Val.*

1.   **Aprion virescens.**   [361.]

Aprion virescens, *Cuv. & Val.* vi. p. 544, pl. 168; *Günth. Fish.* i. p. 81.
  Seychelles.

ANYPERODON, *Günth.*

2.   **Anyperodon leucogrammicus.**   [270.]

Serranus leucogrammicus, *Cuv. & Val.* ii. p. 347.
Anyperodon leucogrammicus, *Günth. Fish.* i. p. 96.
  Zanzibar. Seychelles. *Molucca Sea.*

SERRANUS, *Cuv.**

3.   **Serranus louti.**   [358.]

Perca louti, *Forsk.* p. 40.
Serranus louti, *Rüpp. Atlas, Fische,* p. 106, taf. 26. f. 2 ; *Günth. Fish.* i. p. 101.
  Zanzibar. *Red Sea. Mauritius. Ceylon. Molucca Sea. Sumatra. Timor. Waigiou.*

---

* Doubtful specimens [24, 563].

B

4. **Serranus boelang.** [211.]

Serranus boelang, *Cuv. & Val.* ii. p. 308, vi. p. 514; *Quoy & Gaim. Voy. Astrol. Poiss.* p. 657, pl. 3. fig. 4.

—— nigrofasciatus, *Hombr. & Jaq. Voy. Pôle Sud, Poiss.* p. 36, pl. 2. fig. 1; *Peters, Monatsber. Ak. Wiss. Berl.* 1865, p. 105.

—— bocnak, *Günth. Fish.* i. p. 112 (not *Bloch*).

$$D. \frac{9}{15\text{-}16}. \quad A. \frac{3}{5}. \quad L. \text{ lat. } 70.$$

Caudalis rounded. Posterior limb of præoperculum slightly emarginate and finely denticulated. The height of the body is nearly equal to the length of the head, and is contained three times and a half in the total length. The diameter of the eye equals one-fifth of the length of the head, in a species 6 inches long. The upper maxillary bone reaches beyond the vertical from the posterior margin of the orbit. The pectorals are longer than the ventrals, the former reaching as far as the vent. The dorsal spines, the first three excepted, are subequal. The second of the anal is as long as, but stronger than the third, and two-fifths of the length of the head.

Colour madder-brown, with seven or eight darker cross bands. A still darker spot between the two upper opercular spines. Vertical fins brown, with black edges and scarcely perceptible white marginal lines. Pectorals uniform dark brown.

Zanzibar. *Sunda and Molucca seas.*

5. **Serranus rogaa.** [289, 374.]

Perca rogaa, *Forsk.* p. 38.

Serranus rogaa, *Cuv. & Val.* ii. p. 349; *Rüpp. Atlas, Fische,* p. 105, taf. 26. fig. 1; *Günth. Fish.* i. p. 116.

Var. *a* [374]. Uniform blackish brown; fins black-edged. Length 14 inches.

Var. *b* [289]. Uniform blackish brown; the caudal with a broad white posterior margin. The posterior part of soft dorsal with a narrow white margin. It may, however, only be an immature specimen. Its length is 9 inches.

Zanzibar. *Red Sea.*

6. **Serranus erythræus.** Plate 1. fig. 1. [427.]

Serranus erythræus, *Cuv. & Val.* vi. p. 516; *Günth. Fish.* i. p. 116.

$$D. \frac{9}{15\text{-}16}. \quad A. \frac{3}{9}. \quad L. \text{ lat. } 120.$$

Caudalis rounded. The height of the body is contained thrice and two-thirds, and the length of the head thrice and three-fourths in the total length. The eye is small, and is contained six times and a half in the length of the head. Upper profile of the head concave. The upper maxillary bone has at its lower anterior angle a process pointing forwards. Præoperculum finely serrated on both limbs, and distinctly emarginate above the angle. Sub- and interoperculum conspicuously serrated. Operculum with three very broad flat spines. Dorsal spines increasing in length posteriorly, the last being shorter than the shortest ray, and one-third of the length of the head. Second anal spine rather shorter but much stronger than the third, which is about equal to the last dorsal spine. The soft dorsal is about as high as the tail between the terminations of the vertical fins. Colour in life red (yellowish in a dried state); caudal spotted with lighter.

Seychelles. *Mauritius.*

7. **Serranus cyanostigmatoides.** [55.]

Serranus guttatus, *Cuv. & Val.* ii. p. 357 (not *Peters*).
—— cyanostigmatoides, *Bleek. Verhand. Bat. Genootsch.* 1849, *Perc.* p. 31; *Günth. Fish.* i. p. 117.

Zanzibar. *Seas of Java and Amboyna.*

8. **Serranus miniatus.** [152.]

Perca miniata, *Forsk.* p. 41. no. 41.
Serranus miniatus, *Rüpp. Atlas, Fische,* p. 106, t. 26. f. 3; *Günth. Fish.* i. p. 118.

. Aden. *Red Sea. Mozambique.*

9. **Serranus hemistictus.** [54.]

Serranus hemistictus, *Rüpp. Atlas, Fische,* p. 109, t. 27. f. 3; *Günth. Fish.* i. p. 119.

Aden. *Red Sea.*

10. **Serranus sonnerati.** Plate III. fig. 1. [583.]

Serranus sonnerati, *Cuv. & Val.* ii. p. 299; *Günth. Fish.* i. p. 122.

$$D. \frac{9}{15}. \quad A. \frac{3}{9}. \quad L. \text{ lat. } 85.$$

Caudalis rounded. Height of body nearly equal to the length of the head, and contained thrice and a half in the total length. The diameter of the eye is one-fifth of the length of the head. The upper maxillary reaches beyond the posterior margin of the eye. The denticulations of the præoperculum are exceedingly feeble, the posterior limb slightly emarginate; sub- and interoperculum entire. Dorsal spines short, moderately strong, increasing in length posteriorly, the last being two-sevenths of the length of the head. The second anal spine is scarcely longer and stronger than the third, and is equal in length to the last of the dorsal. The pectorals reach to the anal.

Colour dark red; head, belly, dorsal, and anal fins spotted with vermilion, the two last with black margins.

Zanzibar. *Pondicherry. Ceylon. Sumatra. Louisiade archipelago.*

11.   **Serranus lanceolatus.**   [567.]

Holocentrus lanceolatus, *Bloch*, t. 242. f. 1.
Serranus lanceolatus, *Cuv. & Val.* ii. p. 316; *Günth. Fish.* i. p. 107; *Day's Fishes of Malabar*, tab. i. f. 1 (probably not fig. 2).

It has been maintained that this fish is only the young state of some larger species; and it is undoubtedly true that, so far as we are aware, none of the *banded* specimens of *S. lanceolatus* exceed a length of 6 or 7 inches.

Mr. Blyth was the first to refer *lanceolatus* as a synonym to another species, namely to *S. coioides*, Buch. Ham. (= *S. suillus*, Cuv. & Val.) (Journ. As. Soc. Ben. xxix. p. 111). Mr. Day, without referring to this paper, also represents *lanceolatus* as a young *Serranus*, but takes it to be that of *horridus*, K. & v. H. We may remark at once that, to judge from the figures given by Mr. Day, this appears rather improbable, and he does not explain, or even notice, the difference in the length of the dorsal spines in the two fishes.

The British Museum has received from Captain Mitchell of Madras a specimen, 16 inches long, as the old state of *lanceolatus*, which agrees structurally, though not in coloration, with *S. suillus*; it is certainly not the same as Mr. Day's so-called adult *lanceolatus*.

The first question which presented itself to us was, whether we should be able to recognize one of the original figures given by Russell on plates 127 and 128. They represent two fishes which, to say the least, are very closely allied; Russell himself says that 128 may perhaps be merely a variety of 127. Buchanan Hamilton is of opinion that his *Bola coioides* is identical with plate 128; and Cuvier and Valenciennes establish for plate 127 the specific name of *S. suillus*, while plate 128 is called by them *S. bontoo*.

On comparing the figures and descriptions quoted, we cannot come to any conclusion as to whether they refer to one and the same or to two separate fishes.

In the Zanzibar collection there are, besides specimens of the banded *S. lanceolatus*, larger and smaller ones, from 9 to 20 inches in length, which agree very well with plates 127 and 128 of Russell. We might say that they agree better with the former, inasmuch as the pectoral and anal fins are spotted, whilst those fins are immaculate in the latter; therefore we have not hesitated to name these specimens *S. suillus*.

The second question was whether *suillus* represents the more developed state of *lanceolatus*; and we have come to the conclusion that it does not,—*first*, because the largest specimen of *lanceolatus* is scarcely inferior in size to the smallest of *suillus*; secondly, the opercular spines of *lanceolatus* are much more distant than in *suillus*; and thirdly, the scales of *lanceolatus* are very slightly ctenoid, nearly smooth, whilst those of *suillus* have the margin beset with very distinct spinous teeth. These points of difference apply equally to the specimen sent by Capt. Mitchell from Madras as an adult *lanceolatus*, and which at present we are inclined to regard as a variety of *suillus*.

We may remark that the description given by Valenciennes of *S. horridus* is too indefinite to admit of recognition. A specimen in the British Museum received from Dr. Cantor as "*S. horridus*, juv.," measures 10 inches in length, and is certainly specifically identical with the banded *lanceolatus*: the bands have been broken up into the irregular sinuous markings described by Cantor, the coloration of the fins is the same in both, and so is the comparative length of the dorsal spines, which is certainly not the case in the specimens figured by Mr. Day. We do not venture to say what the specimen stated by Mr. Day to be the old state of *lanceolatus* may eventually prove to be.

Zanzibar. Indian Ocean. Batavia. Samarang.

12. **Serranus suillus.** [514, 742.]

Russell, pl. 127 ; ? pl. 128.
Bola coioides, *Buch. Ham.* pp. 82, 369.
Serranus suillus, *Cuv. & Val.* ii. p. 335 ; *Günth. Fish.* i. p. 127.
?*Russell*, pl. 128.
?Serranus bontoo, *Cuv. Règne Anim.; Cuv. & Val.* ii. p. 334, vi. p. 523 ; *Cantor, Mal. Fish.* p. 11; *Günth. Fish.* i. p. 138.

$$D. \frac{11}{13}. \quad A. \frac{3}{8}.$$

Caudalis rounded ; denticulations of the præoperculum conspicuous, stronger at the angle. Height of the body one-fourth of the total length ; the length of the head is contained thrice and one-third in the same. The upper maxillary bone reaches to the vertical from the posterior margin of the eye. Pectorals rather longer than the ventrals, but not reaching the vent. Dorsal spines, the first two excepted, subequal in length. The second and third of anal equal in length and strength, and two-sevenths of the length of the head.

Colour brownish, with eight more or less interrupted darker cross bands,—the first over the orbit and præoperculum ; the second on the nape of the neck ; the third between the second and third dorsal spines ; the fourth, very indistinct, from the eleventh to the fifteenth spine ; the fifth from the first to the seventh ray ; the sixth from the ninth to the last ray ; the seventh before the caudal ; and the eighth across the root of the caudal. The head and body are moreover covered with large, round, yellowish-brown spots, which are smaller than the interspaces between them.

Zanzibar. Philippines. Java. Coasts of India.

13. **Serranus fuscoguttatus.** [410, 494.]

Percæ summanæ, var. *b, Forsk.* p. 42.
Serranus fuscoguttatus, *Rüpp. Atlas, Fische,* p. 108, t. 27. f. 2 ; *Peters, Wiegm. Arch.* 1855, p. 235 ; *Günth. Fish.* i. p. 127.

$$D. \frac{11}{13-15}. \quad A. \frac{3}{8-9}.$$

Caudalis rounded. Denticulation of the præoperculum stronger at the angle. Sub- and interoperculum entire. The height of the body is contained thrice and three-fourths, and the length of the head thrice and one-fourth in the total length. The diameter of the eye is one-seventh of the length of the head. The pectorals do not reach the vent.

Colour light brown, marbled and spotted with darker. There are also a number of large irregular blackish-brown spots arranged as five or six interrupted cross bands, the last of which crosses the tail behind the dorsal. The snout has a reddish tinge. The fins are marbled and spotted with blackish brown, the interstices having a reddish tinge toward the margin.

Zanzibar. Red Sea. Mozambique. Hope Island, N.E. coast of Australia.

14. **Serranus dispar**, sp. n. Plate I. figs. 2, 3. [423, 389.]

$$D. \frac{11}{14-15}. \quad A. \frac{3}{8}. \quad L. lat. c. 80.$$

Diagnosis.—Caudalis rounded. The height of the body is contained four times, and the length of the head thrice and two-thirds in the total length. The diameter of the eye is contained five times and a half in the length of the head. The upper maxillary bone reaches to the vertical from the posterior margin of the orbit. Upper limb of the præoperculum rather strongly serrated, with a few stronger denticulations at the angle. Sub- and interoperculum entire. The third, fourth, and fifth spines of the dorsal are the longest, and about one-third the length of the head. Pectorals longer than the ventrals.

The two varieties described below differ greatly in coloration; but they have one common feature in the series of large round spots, about the size of the orbit, which runs along the back.

Description.—The form is oblong, rather elongated: the greatest height is above the ventrals; it is one-fourth of the total length, and nearly half the length of the dorsal fin. The length of the head is contained three times and two-thirds in the total length. The distance between the eyes is about equal to their diameter, which is contained five times and a half in the length of the head. The length of the snout is about once and a half the diameter of the eye.

The cleft of the mouth is moderately oblique, the upper maxillary bone reaching the vertical from the posterior margin of the orbit. The upper limb of the præoperculum is rather strongly serrated, with a few stronger denticulations at the angle; the lower limb is scarcely, and the sub- and interoperculum not serrated. The operculum terminates in three spines: the two upper are conspicuous, flat, short, and triangular; the third is hidden in the scales. The lower margin of the upper and the upper margin of the second spine form nearly a right angle.

The lower two-thirds of the dorsal are covered with minute scales. The spinous portion is lower than the soft. The spines are of moderate strength; the third, fourth, and fifth are the longest, and about one-third the length of the head; the second is equal to the last; there is a shallow notch between the spinous and soft portions. The soft portion is subquadrangular in shape, with a curved upper margin. The first ray is one-third longer than the last spine; the third to the twelfth are subequal in length.

The caudalis is rounded, narrow bands of scales running between the rays to three-fourths of the length of the fin. The anal commences a little behind the beginning of the soft dorsal; the first spine is feeble, and one-half the length of the second, which is much stronger; the third is longer, but weaker than the second; it is about equal to the second of the dorsal, and is contained four times in the length of the head. The posterior angle of the soft portion is rounded; the longest rays are the fourth, fifth, and sixth, which equal the longest of the dorsal. The pectorals consist of eighteen rays; they are rounded, longer than the ventrals, and covered with small scales to one-third of their length. The ventrals are rather small, and do not nearly reach the vent; the spine is equal to the second of the dorsal; the longest ray (the third) is longer than the longest of the dorsal.

The canine teeth are very feeble in the upper, and inconspicuous in the lower jaw.

Var. a [423]. The ground-colour is brownish grey, lighter on the belly. On the body there are about five rather irregular obliquely transverse series of large darker spots, about as large as the orbit; those above are rather larger, and those on the lower parts of the body are rather smaller. The upper parts of the head are covered with small brown spots; all the fins, the spinous portion of the dorsal excepted, are covered with rather crowded, round, dark-brown spots, much smaller than those on the body, and larger than those on the head.

A single specimen of this variety was obtained at the Seychelles.

Var. b [389]. Colour light brown, with two longitudinal series of large, rounded, brown spots, about as large as the orbit,—the upper composed of five spots, running along the base of the dorsal; the lower formed of three or four. The head, body, and fins are covered with numerous densely crowded subpolygonal dark-brown spots, each about the size of the pupil; they are most distinct on the lower parts, where they are separated by a network of whitish lines. There is a subquadrangular black blotch on the back of the tail, behind the dorsal.

Two specimens of this variety were found at Zanzibar.

Both varieties appeared to be of about the same age; their length is 15 inches.

15.   **Serranus marginalis.**   [53, 458.]

Epinephelus marginalis, *Bloch*, t. 328. f. 1.
Serranus marginalis, *Cuv. & Val.* ii. p. 301; *Günth. Fish.* i. p. 135; *Peters, Monatsber. Ak. Wiss.*
    Berlin, 1865, p. 109.

Colour in life bright red, with four or five darker cross bands; a darker red patch on the forehead, from the base of the first dorsal spine to the snout, including the eye. Another similarly coloured band from snout above the maxillary to the angle of the præoperculum. The spinous portion of the dorsal, and sometimes the caudal, black-edged; the soft dorsal with a yellowish margin.

Dr. Bleeker and Prof. Peters regard *Perca fasciata* (Forsk.) or *Serranus oceanicus* (Lacép.) merely as a variety of coloration of this species.

Aden.   Zanzibar.   Johanna.   *Javanese, Japanese, and China seas.   Amboyna. Timor.   Louisiade archipelago.*

16.   **Serranus summana.**   Plate II. fig. 1.   [265.]

Perca summana, *Forsk*, p. 42.
Bodianus summana, *Bl. Schn.* p. 334.
Pomacentrus summana, *Lacép.* iii. p. 511.
    But not
    Serranus summana, *Cuv. & Val., Rüpp., & Lefèbvre.*

$$D. \frac{11}{15\text{--}16}. \quad A. \frac{3}{8}.$$

Caudalis rounded. Canine teeth very small. Præoperculum distinctly serrated, with a very shallow notch above the angle. The third to the sixth dorsal spines are the longest, rather feeble, not quite one-third the length of the head. The height of the body is contained thrice and two-thirds, and the length of the head three times in the total length. The diameter of the eye is one-sixth of the length of the head. The upper maxillary bone reaches beyond a vertical from the posterior margin of the orbit.

Head, body, and fins brown; the trunk and vertical fins covered with very small, round, white dots, each being smaller than a scale. The head is nearly immaculate. There is a blackish streak above the maxillary bone.

Note.—This is clearly the *S. summana* of Forskal, who describes it as covered with *guttæ*, not with *maculæ*, and with hardly any on the head. There are two specimens in the British Museum, received from Dr. Rüppell, with the name of *S. summana*; but these are identical with *S. tumilabris* of this collection [122].

Zanzibar.   *Red Sea.*

17.   **Serranus tumilabris.**   Plate II. fig. 2.   [122.]

Serranus summana, *Cuv. & Val.* ii. p. 344; *Rüpp. N. W. Fische*, p. 102, and *Atlas*, p. 104; *Lefèbvre, Voy. Abyss. Zool.* p. 329, pl. 5. fig. 1 (bad).
Serranus tumilabris, *Cuv. & Val.* ii. p. 346; *Günth. Fish.* i. p. 138.

$$D. \frac{11}{15}. \quad A. \frac{3}{8}.$$

Caudalis rounded. The height of the body is contained thrice and a half, and the length of the head thrice and one-seventh in the total length. The diameter of the eye is one-fifth of the length of the head. Snout rather pointed, longer than the eye. Præoperculum slightly emarginate behind, conspicuously denticulated. Sub- and interoperculum entire. The maxillary bone reaches nearly to a vertical from the posterior margin of the orbit. Dorsal spines feeble, subequal in length, about one-third the length of the head. The second and third of the anal are equal; the former is rather the stronger.

Colour brown, covered all over with round white spots. In addition there are five or six obliquely transverse rows of larger, light-coloured spots, through which the smaller ones are distinctly visible. The greatest number of the larger spots in a row is four, and each contains from three to nine of the smaller; the foremost row is on the head. All the fins are spotted with white, each spot being as large as the pupil. There is a black streak above the maxillary bone.

Aden. *Red Sea.*

18.   **Serranus hoevenii.** Plate II. fig. 3.   [306, 422, 600, 711.]

Serranus hoevenii, *Bleek. Verhand. Bat. Genootsch.* 1849, *Perc.* p. 36; *Günth. Fish.* i. p. 138.

$$D. \frac{11}{16}. \quad A. \frac{3}{8}. \quad L. \text{lat.} 80.$$

Caudalis rounded. The height of the body is contained thrice and three-fourths, and the length of the head thrice and one-fourth in the total length. Upper profile of the head nearly straight, muzzle pointed. The diameter of the eye is one-sixth of the length of the head, and a little less than that of the snout. Præoperculum distinctly serrated behind, scarcely emarginate above the angle. Sub- and interoperculum entire. The upper maxillary bone does not quite reach as far as the posterior margin of the eye. The dorsal spines are of moderate strength, subequal in length, the first two excepted, the longest being one-third the length of the head. The soft is higher than the spinous portion.

Body and dorsal blackish, covered all over with larger and smaller round pearly spots; the larger ones scarcely exceed the pupil in size. The pectorals, ventrals, caudal, and anal black, the last with a few white spots on the base. A black streak above the maxillary.

This is undoubtedly the fish described by Bleeker, from whose collection a specimen was received at the British Museum. He describes the fins as immaculate; but small pearly spots are clearly discernible on the posterior part of the soft dorsal, the base of the anal, and the caudal. We have examined five specimens from Zanzibar, and we find that the size and number of the spots on the fins vary greatly with the age of the fish. On one about the same size as that received from Amboyna, the spots on the fins are nearly as indistinct.

Zanzibar. Mombassa. *Amboyna.*

C

19.   **Serranus hexagonatus.**   [373, 709.]

Perca hexagonata, *Forster, Descr. Anim.* ed. *Licht.* p. 189.
Serranus hexagonatus, *Cuv. & Val.* ii. p. 330; *Günth. Fish.* i. p. 141; *Peters, Monatsber. Ak. Wiss.
    Berlin,* 1865, p. 110; *Kner, Novara, Fisch.* p. 25*.
Serranus merra, *Cuv. & Val.* ii. p. 325.

Var. *a* [373]. Brownish, covered all over with hexagonal spots. Length 5 inches.
Zanzibar.

Var. *b* [373]. The spots are hexagonal, four larger darker blotches, three of which
are on the base of the dorsal, and one on the back of the tail; the first is the largest,
and is situated below the last dorsal spine. Length 6 inches.
Great Comoro.

Var. *c* [373]. With subhexagonal and rounded spots. Length 5 inches.
Zanzibar. Great Comoro.

Var. *d* [709]. With alternate cross bands of lighter and darker spots. Length
8 inches.
Zanzibar.

20.   **Serranus longispinis.**   [550.]

? Serranus gaimardi, *Bleek. Nat. Tydschr. Ned. Ind.* 1853, *Batav.* p. 455; *?? Quoy & Gaim. Voy.
    Astrol. Poiss.* p. 656, pl. 3. f. 3; *?? Günth. Fish.* i. p. 150.
Serranus longispinis, *Kner, Novara, Fische,* p. 27, t. 2. f. 2.

$$\text{D. } \frac{11}{14-17}. \quad \text{A. } \frac{3}{8}. \quad \text{L. lat. 100.}$$

The third and fourth dorsal spines longest, and equal to half the length of the head.
Præoperculum emarginate above the angle; at it the denticulations are considerably
stronger. Operculum with two spines. Caudal truncated. Head, body, and all the fins
with numerous brown spots. The anterior part of the dorsal with a black margin.
Zanzibar. *Madras.*

21.   **Serranus flavo-cæruleus.**   [268, 269.]

Holocentrus flavo-cæruleus, *Lacép.* iv. p. 366.
Serranus flavo-cæruleus, *Cuv. & Val.* ii. p. 297; *Günth. Fish.* i. 145.

---

* Prof. Kner says that he is unable to find the distinction between this species and *S. gilberti.* If he had
read with attention the descriptions given, he would have found that the latter *is* distinguished by much longer
pectoral fins.

Some specimens have the tip of each caudal lobe and of the ventrals black, and the base of the dorsal, anal, and pectorals blackish brown.

Zanzibar. *Mozambique. Mauritius. Ceylon.*

22.   Serranus areolatus.   [159, 186.]

Perca arcolata, *Forsk.* p. 42.
Perca tauvina, *Geoffr. Descr. de l'Ég.* pl. 20. fig. 1 (not *Forsk.*).
Serranus arcolatus, *Schleg. Fauna Japon.* p. 8; *Cuv. & Val.* ii. p. 350; *Günth. Fish.* i. p. 149.

Zanzibar. Aden. *Red Sea. Mozambique. Sea of Japan.*

23.   Serranus cylindricus.   [373 A.]

Serranus cylindricus, *Günth. Fish.* i. p. 151, plate xi. f. A.

Zanzibar. *Madagascar.*

24.   Serranus striolatus, sp. n.   Plate III. fig. 2.   [359.]

$$D. \frac{11}{13}. \quad A. \frac{3}{10}.$$

Diagnosis.—Caudal rounded, the height of the body is contained thrice and one-third in the total length, the length of the head four times and one-sixth in the same. Diameter of the eye one-seventh of the length of the head. The upper maxillary bone reaches to the vertical from posterior margin of orbit. Præoperculum inconspicuously serrated behind, the denticulation rather stronger at the angle. Sub- and interoperculum entire. The third and fourth dorsal spines are longest and one-third the length of the head. General ground-colour brownish yellow. Body with about fifteen more or less undulated series of short brown streaks, which are sometimes confluent. Head with numerous small round brown spots. The head and trunk show, besides, numerous rounded lighter spots, as in *S. tumilabris*, some of which have a brown spot in the centre.

Description.—This species somewhat resembles *S. altivelis*. The neck is elongated, the head strongly compressed and high, with its upper profile concave. The body is elevated, its greatest height being below the origin of the dorsal, where it is contained thrice and one-third in the total length. The length of the head is contained four times and one-sixth in the same. The distance between the eyes is much less than their diameter, and is covered with minute scales as far as the nostrils. The eye is small, being one-seventh of the length of the head, and half that of the snout. The upper maxillary bone reaches to the vertical from the posterior margin of the orbit. The præoperculum is inconspicuously serrated behind, the denticulation being rather stronger at the angle. The sub- and interoperculum are entire. The lips are thick and fleshy. The operculum terminates in three feeble inconspicuous spines, the middle one being more remote from the first than from the third.

c 2

The basal portion of the dorsal is scaly; the spines are strong, the third and fourth are the longest, their length being more than one-third of that of the head. The margin of the spinous portion is concave behind, the eighth spine being a little shorter than the last three. The soft dorsal is much elevated, the longest (eleventh) ray being half the length of the head. Caudal rounded posteriorly. The anal is as much elevated as the soft dorsal; the spines are short and stout, the third being rather longer than the second, and about one-fourth of the length of the head. The teeth, and especially the canines, are very feeble.

Length 19 inches.

Zanzibar.

PLECTROPOMA, *Cuv.*

25.   **Plectropoma maculatum.**   [316, 421, 444, 474, 566.]

This species is subject to very great diversities of colour, without the slightest structural divergence. We have examined seven distinct varieties, one of which (*d*) has been described as *P. maculatum*, and another (*a*) as *P. melanoleucum*; the other five appear to be hitherto unnoticed.

The specimens examined form a very complete series; each variety exhibits strong points of similarity to those before and after it, while the extremes are so different that, without an examination of the intermediate varieties, they might well be regarded as distinct species. We regret much that our series is chiefly composed of dried specimens; and it must be left to future researches to ascertain whether sex or season has any influence on the coloration of this fish.

Synonymy of var. *d* :—

Bodianus maculatus, *Bloch*, t. 228 ; *Lacép.* iv. pp. 280, 293.

Plectropoma punctatum, *Quoy & Gaim. Voy. Freycin. Zool. Poiss.* p. 318, t. 45. f. 1.

—— maculatum, *Cuv. & Val.* ll. p. 330, *Bleek. Verhand. Bat. Genootsch.* xxii. (1849) *Perc.* p. 39, and (1850) *Jav.* p. 418; *Günth. Fish.* i. p. 156.

Synonymy of var. *a* :—

Renard, i. pl. 22. f. 120; *Valentyn*, f. 497.

Bodianus melanoleucus, *Lacép.* iv. pp. 283, 297.

—— cyclostoma, *Lacép.* iii. pl. 20. f. 1, iv. pp. 282, 293.

Labrus lævis, *Lacép.* iii. pp. 431, 479, pl. 23. f. 2.

Plectropoma melanoleucum, *Cuv. & Val.* ii. p. 388; *Peters, Wiegm. Arch.* 1855, p. 238; *Günth. Fish.* i. p. 157.

$$\text{D.}\ \frac{7\text{-}8}{10\text{-}12}.\quad \text{A.}\ \frac{3}{7\text{-}8}.\quad \text{L. lat. 110.}$$

Diagnosis.—Præoperculum exceedingly finely serrated (or entire) behind, with from three to five spinous teeth below. Sub- and interoperculum entire. Caudalis truncated.

The third to the fifth dorsal spines longest; the spinous portion much lower than the soft or than the anal. The first spine of the anal is very small and frequently concealed in the skin, but never absent as stated by Cuvier and Valenciennes in their description of *P. melanoleucum.*

Var. *a. P. melanoleucum* [316]. Colour violet grey; snout brown; posterior part of tail and point of lower lip yellow. Body with five black cross bands: the first across the occiput from the middle of the interspace between the eyes; the second over the shoulders to the posterior limb of the operculum; the third from before the first to the fifth dorsal spines, continued on that fin and joining the band on the opposite side round the belly; the fourth below the last three dorsal spines; and the fifth from the fifth to last dorsal rays; the last two bands are continued on the base of the dorsal. All the fins are bright yellow; the pectorals and ventrals have large black spots on the base: this last is characteristic of all the varieties.

Length 13½ inches.

Zanzibar. *Mozambique. Mauritius.*

Var. *b* [316]. Similar in general appearance to *a.* The ground-colour is darker; the cross bands have lost their deep-black colour, and have assumed a greenish hue, and their margins *only* are closely studded with small blue dark-edged ocelli.

Length 15 inches.

Zanzibar.

Var. *c* [474]. The ground-colour is nearly similar to that of *b.* A trace still remains of the cross bands, but it is so faint that it only becomes apparent in a dried state and in certain lights. The *whole* of the head, body, vertical fins, and the base of the pectorals have become covered with ocelli, as in *b.* They are small and round, each occupying about three scales.

This specimen is about the same size as the last.

Zanzibar.

Var. *d. P. maculatum* [444]. The entire body has assumed a lighter hue, the ground-colour is yellowish brown, and the ocelli have changed into pale-blue spots, some of which, especially those on the middle of the body and tail, are elongated and streak-like. The bases of the soft dorsal and anal are spotted like the body; the head and caudal have numerous dots of the same colour. This variety has black spots between the first five spines of the dorsal, which are more visible in young than in old specimens. Specimens of this variety are usually much smaller than those of the preceding ones.

Zanzibar. *Red Sea. East-Indian Archipelago.*

Var. *e* [444 *b*]. The ground-colour of the body is darker than in *d*, the spots on the body and fins are far less numerous, and the black spots between the dorsal spines have become extremely indistinct.

Zanzibar.

Var. *f* [566]. The ground-colour has become of a deep brown, the spots have entirely disappeared, but a trace of the cross bands, as in *a* and *b*, is perceptible in a dried state.

Length 11 inches.

Zanzibar.

Var. *g* [421]. Also without spots. Ground-colour brownish, clouded with lighter on the back.

Length 13 inches.

Seychelles.

GRAMMISTES, *Cuv.*

26.   Grammistes orientalis.   [431.]

Grammistes orientalis, *Bl. Schn.* p. 189; *Cuv. & Val.* ii. p. 203; *Günth. Fish.* i. p. 171.

Zanzibar. *Mauritius. East-Indian seas. Australia.*

GENYOROGE, *Cantor* *.

27.   Genyoroge macolor.   [360. 590.]

Synonymy of var. *a. G. macolor* [360]:—

Macolor, *Renard*, i. pl. 9. f. 60, pl. 7. f. 30; *Valent.* iii. p. 348, pl. 1. f. 1.
Diacope macolor, *Cuv. & Val.* ii. p. 415; *Less. Mém. Soc. Hist. Nat.* iv. p. 409, and *Voy. Coq. Zool.* ii. p. 230, pl. 22. f. 2.
Mesoprion macolor, *Bleek. Nat. Tydschr. Ned. Ind.* 1852, *Celebes*, iii. p. 752.
Genyoroge macolor, *Günth. Fish.* i. p. 170.

Synonymy of var. *b. G. nigra* [590]:—

Sciæna nigra, *Forsk.* p. 47.
Lutjanus niger, *Bl. Schn.* p. 326.
Diacope nigra, *Cuv. & Val.* ii. p. 431; *Rüpp. N. W. Fische*, p. 93, t. 24. f. 1.
Genyoroge nigra, *Günth. Fish.* i. p. 176.

$$\text{D.} \frac{10}{13\text{-}13}. \quad \text{A.} \frac{3}{10\text{-}11}.$$

The specimens described by various authors as *G. macolor* and *nigra* are evidently the same fish at different stages of growth. We have examined a series of specimens from Zanzibar, in which the transition from *macolor* to *nigra* is clearly traced. The

* Undetermined specimen [206].

first has the spots on the back and sides well marked; the white longitudinal bands on the body, the posterior cross band on the head, and the white of the belly are hardly separated from each other by the black ground-colour. The second specimen has these markings well defined and separate from each other. In the fourth they have almost, and in the fifth they have entirely disappeared, so that the fish has become uniformly black.

The structure of these specimens is identical, except that the serrature of the præoperculum becomes more strongly marked, the groove deepens, and the spine of the interoperculum becomes larger with age.

Zanzibar. Aden. *Red Sea. New Guinea. Amboyna. Celebes.*

28.   Genyoroge sebæ.   [97.]

Seba, iii. 27. 2; *Russell*, pl. 99.
Diacope sebæ, *Cuv. & Val.* ii. p. 411.
—— siamensis, *Cuv. & Val.* vi. p. 524.
Genyoroge sebæ, *Günth. Fish.* i. p. 176.

Aden. Zanzibar. *Mozambique. Coast of India. Seas of Java, Amboyna, Waigiou. Louisiade archipelago.*

29.   Genyoroge bengalensis.   [28, 342.]

Sciæna kasmira, *Forsk.* p. 46.
Holocentrus bengalensis, *Bloch*, t. 246. f. 2.
Labrus octolineatus, *Lacép.* iv. p. 478, pl. 22. f. 1.
Diacope octolineata, *Cuv. & Val.* ii. p. 418, vi. p. 526; *Schl. Faun. Japon.* p. 12, pl. 6. f. 2.
Genyoroge bengalensis, *Günth. Fish.* i. p. 178.

Aden. Zanzibar. *Red Sea. East-Indian seas. Mauritius. Polynesia.*

30.   Genyoroge notata.   [235.]

Russell, pl. 98.
Diacope notata, *Cuv. & Val.* ii. p. 422.
Genyoroge notata, *Günth. Fish.* i. p. 181.

Prof. Kner (Novara, Fische, p. 35) supposes that this species is identical with *Mesoprion fulviflamma*, but at the same time thinks it possible that he never saw a true *Genyoroge notata*. As both species are found on the coast of Zanzibar, we can confidently state that they are most distinct species. Besides the difference in the development of the interopercular knob and the corresponding notch on the operculum, *G. notata* has the black spot *above*, while *M. fulviflamma* has it *on* the lateral line; the former also has the second anal spine longer and stronger than the third, its length being contained twice and a quarter in that of the head, while in the latter it is *not* longer and stronger than the third, and its length is one-fifth of that of the head. Furthermore, in *G. notata* the lines on the body are very distinct and of a bright blue colour, while in *M. fulviflamma* they are much less distinct and of a yellowish colour. Russell's figure of *G. notata* and Rüppell's figure of *M. fulviflamma* are sufficiently good.

We may also remark on this occasion that we have convinced ourselves, from an examination of Cantor's typical specimen of *G. notata*, which has been transferred to the British Museum since the publication of the first volume of Dr. Günther's Catalogue, that the fish described by Cantor under that name is not a *Genyoroge*, but a *Mesoprion*, and perhaps identical with *M. russellii* of Bleeker.

Zanzibar, *Bay of Bengal. Penang.*

---

31. **Genyoroge marginata.** [3, 417.]

Diacope marginata, *Cuv. & Val.* ii. p. 425; *Peters, Wiegm. Archiv,* 1855, p. 238.
—— xanthopus, *Cuv. & Val.* iii. p. 495.
Genyoroge marginata, *Günth. Fish.* i. p. 181.
Mesoprion marginatus, *Bleek. Nat. Tydschr. Ned. Ind.* 1852, *Amboina,* ii. p. 554; *Kner, Novara, Fisch.* p. 31.

Zanzibar. *Mozambique. Louisiade archipelago. Amboyna. Ceylon.*

---

32. **Genyoroge bottonensis.** [337, 632.]

Holocentrus boutton, *Lacép.* iv. pp. 331, 367.
Diacope bottonensis, *Cuv. & Val.* ii. p. 434, vi. p. 535.
Genyoroge bottonensis, *Günth. Fish.* i. p. 181.
Mesoprion bottonensis, *Bleek. Nat. Tydschr. Ned. Ind.* ii. p. 170; *Kner, Novara, Fisch.* p. 32, taf. 2. fig. 3.

$$D. \frac{10}{14}. \quad A. \frac{3}{8}. \quad L. \text{ lat. } 55.$$

The specimens of this collection marked [337] and [632] correspond to the specimens in the British Museum from Amboina. In none of these are there any traces of oblique darker *bands*; but in specimens in spirits there are faint *streaks* following the series of scales, which are not apparent either in the living fish or in stuffed specimens.

Zanzibar. *Sumatra. Amboyna. New Guinea.*

---

33. **Genyoroge rivulata.** [87.

Diacope rivulata, *Cuv. & Val.* ii. p. 414, pl. 38; *Rüpp. N. W. Fische,* p. 94.
—— cæruleopunctata, *Cuv. & Val.* ii. p. 424.
Genyoroge rivulata, *Günth. Fish.* i. p. 182.
—— cæruleopunctata, *Günth. Fish.* i. p. 182.

Aden. Zanzibar. *Red Sea. Coast of India. Chinese and Javan seas.*

MESOPRION, *Cuv.* *

34. <div align="center">**Mesoprion bohar.**</div> [326, 366.]

Sciæna bohar, *Forsk.* p. 46.
Diacope bohar, *Cuv. & Val.* ii. p. 433; *Rüpp. Atlas, Fische,* p. 73, and *N. W. Fische,* p. 103.
Diacope quadriguttata, *Cuv. & Val.* ii. p. 427, vi. p. 533.
Mesoprion bohar, *Günth. Fish.* i. p. 190.

Zanzibar. *Red Sea. Madagascar.*

35. <div align="center">**Mesoprion gembra.**</div> [157.]

Russell, pl. 95.
Mesoprion gembra, *Cuv. & Val.* ii. p. 485; *Cantor, Mal. Fish.* p. 15; *Günth. Fish.* i. p. 193.

Zanzibar. *Bay of Bengal. Molucca Sea.*

36. <div align="center">**Mesoprion fulviflamma.**</div> [29.]

Sciæna fulviflamma, *Forsk.* p. 45.
Diacope fulviflamma, *Cuv. & Val.* ii. p. 423; *Rüpp. Atlas, Fische,* p. 72, t. 19. f. 2, and *N. W. Fische,* p. 94.
Mesoprion fulviflamma, *Günth. Fish.* i. p. 201.

Aden. Zanzibar. Scychelles. *Red Sea. Seas of Java, Amboyna, and China.*

37. <div align="center">**Mesoprion annularis.**</div> [271.]

Mesoprion annularis, *Cuv. & Val.* ii. p. 484, iii. p. 497; *Günth. Fish.* i. p. 204; *Kner, Novara, Fisch.* p. 33.
Diacope annularis, *Rüpp. Atlas, Fische,* p. 14, and *N. W. Fische,* p. 91.

Mozambique. *Indian Ocean. Red Sea. Chinese and Japanese seas.*

38. <div align="center">**Mesoprion lineolatus.**</div> [486.]

Diacope lineolata, *Rüpp. Atlas, Fische,* p. 76, t. 19. f. 3.
Mesoprion lineolatus, *Günth. Fish.* i. p. 205; *Kner, l. c.* p. 36.

Var. *a.* Body yellow, with oblique pale-blue streaks above the lateral line, and about seven longitudinal ones below it.

Var. *b.* Olive-brown, the oblique and longitudinal lines darker blue.

Var. *c.* Violet above, muzzle rosy, the oblique and longitudinal lines yellow.

Zanzibar. *Red Sea. Amboyna.*

* Undetermined specimen [694].

D

39. **Mesoprion vitta.** [704.]

Serranus vitta, *Quoy & Gaim. Voy. de Freyc. Zool. Poiss.* p. 315, pl. 58. f. 3.
Diacope vitta, *Schleg. & Temm. Faun. Japon.* p. 13, pl. 6. f. 1.
Mesoprion vitta, *Günth. Fish.* i. p. 207.

Seychelles. *Japanese, Javan, and China seas. Amboyna. Louisiade archipelago. Waigiou. North coast of Australia.*

40. **Mesoprion cæruleo-lineatus.** [21.]

Diacope cæruleo-lineata, *Rüpp. N. W. Fische*, p. 93, t. 24. f. 3 [not *M. quinquelineatus, Cuv. & Val.*].
Aden. *Red Sea.*

PRIACANTHUS, *Cuv. & Val.*

41. **Priacanthus blochii.** [? 58, 678.]

Anthias macrophthalmus, *Bl.* vi. p. 115, t. 319; *Bl. Schn.* p. 304.
Priacanthus japonicus, *Bleeker, Nat. Tydschr. Ned. Ind.* ii. p. 171 (not *Cuv. & Val.*).
—— blochii, *Bleek. l. c.* iv. p. 456; *Günth. Fish.* i. p. 218.
Aden. Mozambique Channel. *Amboyna. Sumatra.*

42. **Priacanthus hamruhr.** [267.]

Sciæna hamruhr, *Forsk.* p. 45.
Priacanthus hamruhr, *Cuv. & Val.* iii. p. 104; *Rüpp. N. W. Fische*, p. 95; *Günth. Fish.* i. p. 219.
Zanzibar. *Red Sea.*

AMBASSIS, *Commers.*

43. **Ambassis commersonii.** [502, 653.]

Sciæna safgha, *Forsk.* p. 53.
Centropomus ambassis, *Lacép.* iv. p. 273.
Ambassis commersonii, *Cuv. & Val.* ii. p. 176, pl. 25; *Rüpp. N. W. Fische*, p. 89; *Günth. Fish.* i. p. 223.

Pangani River, east coast of Africa. *Red Sea. Mauritius. East-Indian seas. North of Australia.*

44. **Ambassis urotænia.** [705.]

Ambassis urotænia, *Bleeker, Nat. Tydschr. Ned. Ind.* 1852, *Amboina & Ceram*, p. 25.
Freshwater of Seychelles. *Seas of Amboyna and Wahai.*

45. **Ambassis dussumieri.** [586.]

Ambassis dussumicri, *Cuv. & Val.* ii. p. 181, vi. p. 503, ix. p. 431 ; *Quoy & Gaim. Voy. Astrol. Poiss.* p. 650, pl. i. f. 3 ; *Günth. Fish.* i. p. 225.

Chanda dussumicri, *Cantor, Mal. Fish.* p. 6.

Zanzibar. *Mauritius. Seychelles. Malabar. Penang. Java. Celebes. Amboyna. China.*

APOGON*, *Lacép.*

46. **Apogon hyalosoma.** [706.]

Apogon thcrmalis, *Bleek. Verh. Bat. Genootsch.* xxii. 1849, *Perc.* p. 27.

—— hyalosoma, *Bleek. Nat. Tydschr. Ned. Ind.* 1852, *Singapore*, p. 63, and *ibid.* 1853, *Amboina*, iv. p. 329 ; *Günth. Fish.* i. p. 231.

Freshwater of Seychelles. *Seas of Batavia, Sumbawa, Amboyna, and Sumatra.*

47. **Apogon nigripes, sp. n.** Plate V. fig. 1. [579.]

D. 6 | $\frac{1}{8}$. A. $\frac{2}{8}$. Br. 7. L. lat. 23. L. transv. 2/5.

Body much elevated, its height being one-half of the total length without caudal. The upper profile descends very rapidly from dorsal towards snout. The head is large, and is contained twice and two-thirds in the total length without caudal. Eye large, its diameter much longer than snout, and more than a third of the length of the head. The lower jaw projects slightly beyond the upper. The maxillary extends to below the centre of the eye. The præoperculum is minutely serrated behind, but its inner edge is entire. Dorsal spines strong; the second rather longer than the third, and as long as the length of the head without snout. Caudal emarginate. Anal spines short; the first is two-fifths, the second three-fifths the length of the second dorsal spine. The first and second dorsal rays much produced, extending to the middle of the anal. The free portion of tail low, nearly twice as long as the eye.

Colour.—Yellowish brown, with numerous irregular vertical silvery cross bands. Throat bright yellow; *ventrals blackish brown.* First and second anal rays, anterior part of second dorsal, and upper and lower caudal lobes brownish.

Length 27 lines.

Zanzibar.

48. **Apogon amboinensis.** [616.]

Apogon amboinensis, *Bleeker, Nat. Tydschr. Ned. Ind.* 1853, *Amboyna*, iv. p. 329 ; *Günth. Fish.* i. p. 234.

Saltwater of Zanzibar. *Rivers of Amboyna.*

* Undetermined specimens [73, 481].

D 2

49. **Apogon nigripinnis.** [629.]

Apogon nigripinnis, *Cuv. & Val.* ii. p. 152; *Günth. Fish.* i. p. 235.

Zanzibar. *Chinese and Japanese seas.*

50. **Apogon tæniopterus.** [525.]

Apogon tæniopterus, *Bennett, Proc. Zool. Soc.* iii. 1835, p. 206; *Günth. Fish.* i. p. 235.

Zanzibar. *Mauritius.*

51. **Apogon bifasciatus.** [480.]

Apogon bifasciatus, *Rüpp. N. W. Fische,* p. 86, t. 22. f. 2; *Günth. Fish.* i. p. 238; *Kner, Novara, Fisch.* p. 42.

Zanzibar. *Red Sea. Chinese seas.*

52. **Apogon annularis.** [237.]

Apogon annularis, *Rüpp. Altas,* p. 48, and *N. IV. Fische,* p. 85.
—— roscipinnis, *Cuv. & Val.* iii. p. 490, vi. p. 553; *Peters, Wiegm. Arch.* 1855, p. 234; *Quoy & Gaim. Voy. Astrol. Poiss.* p. 649, pl. 1. f. 5; *Günth. Fish.* i. p. 239.

Zanzibar. *Mozambique. Red Sea. Indian Ocean.*

53. **Apogon fasciatus.** [202.]

Mullus fasciatus, *White, New South Wales,* p. 268. f. 1.
Apogon novem-fasciatus, *Cuv. & Val.* ii. p. 154; *Peters, Wiegm. Arch.* 1855, p. 234.
—— fasciatus, *Quoy & Gaim. Voy. Freyc. Zool.* p. 344; *Günth. Fish.* i. p. 241.

Zanzibar. Johanna. *Mozambique. Feejee Islands. Molucca seas.*

54. **Apogon cyanosoma.** [676.]

Apogon cyanosoma, *Bleeker, Nat. Tydschr. Ned. Ind.* 1853, *Solor,* p. 71; *Günth. Fish.* i. p. 242.

Zanzibar. *Solor.*

55. **Apogon macropteroides.** [307, 573, 622.]

Apogon macropteroides, *Bleeker, Nat. Tydschr. Ned. Ind.* 1852, *Banka,* ii. p. 724; *Günth. Fish.* i. p. 245.

$$\text{D. } 6 \mid \tfrac{1}{9}. \quad \text{A. } \tfrac{2}{14\text{-}16}. \quad \text{L. lat. } 23.$$

Var. *a* [573]. Colour rosy, minutely dotted with black, the points on the sides of the head being largest; those at the root of the caudal are crowded, and form a blackish spot. Body with numerous interrupted red transverse lines. Chin bright yellow, a bright yellow band passing over the muzzle, above the upper lip, to the posterior of the orbit.

Length 3 inches.

Zanzibar. *Sea of Japan.*

Var. *b* [307]. Colour silvery, with a yellowish tint, dotted like var. *a*. A yellow line from snout, above eye, terminating below soft dorsal. A series of golden spots, sometimes confluent posteriorly, from above angle of operculum to root of caudal; a few fainter spots and short lines below.

Length 3¾ inches.

Zanzibar.

Var. *c* [622]. Colour whitish, dotted as varieties *a* and *b*, but more sparingly. Tip of first dorsal blackish.

Length 3¼ inches.

Zanzibar.

APOGONICHTHYS, *Bleek*.

56. **Apogonichthys auritus.** [526, 623.]

Apogon auritus, *Cuv. & Val.* vii. p. 413.
—— punctulatus, *Rüpp. N. W. Fische*, p. 88, t. 22. f. 4 (not *Bleek.*).
Apogonichthys auritus, *Günth. Fish.* i. p. 246.

Immature specimens [623] have no opercular spot, or it is merely indicated by an assemblage of minute black punctulations.

Zanzibar. *Mauritius. Red Sea.*

CHILODIPTERUS, *Lacép.*

57. **Chilodipterus octovittatus.** [456.]

Cheilodipterus lineatus, *Lacép.* iii. p. 543, pl. 34. f. 1.
Chilodipterus octovittatus, *Cuv. & Val.* ii. p. 163; *Günth. Fish.* i. p. 248.

Zanzibar. Johanna. *Red Sea. Indian Ocean.*

58. **Chilodipterus lineatus.** [44, 581.]

Perca lineata, *Forsk.* p. 42; *Rüpp. N. W. Fische*, p. 89.
—— arabica, *Linn. Syst. Nat.* i. p. 1312.
Cheilodipterus arabicus, *Cuv. & Val.* ii. p. 165, pl. 23.
Chilodipterus lineatus, *Günth. Fish.* i. p. 248.

Zanzibar. Aden. *Red Sea. Madagascar.*

59. **Chilodipterus quinquelineatus.** [76, 297.]

Cheilodipterus quinquelineatus, *Cuv. & Val.* ii. p. 177; *Rüpp. N. W. Fische,* p. 89; *Günth. Fish.* i. p. 248.

Apogon novemstriatus, *Rüpp. N. W. Fische,* p. 85, t. 22. f. 1.

Zanzibar. Aden. *Red Sea. Amboyna. Society Islands.*

DULES, *Cuv. & Val.*

60. **Dules fuscus.** [339.]

Dules fuscus, *Cuv. & Val.* iii. p. 118; *Peters, Wiegm. Arch.* 1855, p. 238; *Günth.* i. p. 268.

Johanna. *Mauritius. Réunion.*

Family PRISTIPOMATIDÆ.

THERAPON, *Cuv.*

61. **Therapon theraps.** [263.]

Therapon theraps, *Cuv. & Val.* iii. p. 129, pl. 53; *Rüpp. N. W. Fische,* p. 95; *Günth. Fish.* i. p. 274.

Zanzibar. *Indian and China seas; False Bay, entering rivers.*

62. **Therapon servus.** [89.]

Sciæna jarbua, *Forsk* p. 50.

Holocentrus servus, *Bloch,* t. 238. f. 1.

Therapon timoriensis, *Quoy & Gaim. Voy. Uran. Poiss.* p. 341.

—— servus, *Cuv. & Val.* iii. p. 125, vii. p. 479; *Rüpp. N. W. Fische,* p. 95; *Günth. Fish.* i. p. 278.

Aden. Zanzibar. *Red Sea. Indian Ocean to north coast of Australia, entering rivers.*

63. **Therapon trivittatus.** [475.]

Russell, ii. pl. 126.

Coius trivittatus, *Buch. Ham.* pp. 92, 370.

Therapon puta, *Cuv. & Val.* iii. p. 131.

—— trivittatus, *Cantor, Mal. Fish.* p. 19; *Günth. Fish.* i. p. 280; *Kner, Novara, Fisch.* p. 45.

Zanzibar. *Indian seas.*

64.   **Therapon cuvieri.**   [27, 283.]

Pristipoma sexlineatum, *Quoy & Gaim. Voy. Freyc. Poiss.* p. 320.
Pelates sexlineatus, quadrilineatus, *et* quinquelineatus, *Cuv. & Val.* iii. p. 146, pl. 55.
Therapon cuvieri, *Bleek. Nat. Tydschr. Ned. Ind.* vi. p. 211; *Günth. Fish.* i. p. 282.

Zanzibar.   Aden.   *Coasts of Australia.   Sea of Timor.*

PRISTIPOMA, *Cuv.*

65.   **Pristipoma hasta.**   [651A.]

Lutjanus hasta, *Bloch*, t. 246. f. 1.
Labrus commersonii, *Lacép.* iii. pp. 431, 477, pl. 23. f. 1.
Pristipoma kakaan, *Cuv. & Val.* v. p. 244; *Rüpp. N. W. Fische*, p. 123, t. 20. f. 1.
—— commersonii, *Cuv. & Val.* v. p. 252; *Cantor, Mal. Fish.* p. 72.
—— hasta, *Cuv. & Val.* v. p. 247; *Günth. Fish.* i. 289.

Bagamoia, east coast of Africa. *Red Sea. Indian Ocean. North coast of Australia.*

66.   **Pristipoma multimaculatum, sp. n.**   Plate III. fig. 3.   [652.]

D. 11 | $\frac{1}{13}$, A. $\frac{3}{7}$, I. lat. 51. L. transv. $\frac{7-8}{16}$.

Diagnosis.—Height of body equals length of head, and is contained thrice and two-thirds to four times in the total length. Snout longer than the diameter of the eye, which is contained four times and a half in the length of the head. The upper maxillary does not reach the anterior of the eye. Præorbital scaly. Posterior limb of præoperculum emarginate, with the angle rounded and more strongly denticulated. The fourth dorsal spine longest, and more than half the length of the head. Dorsal deeply notched; caudalis truncated. The second anal spine is very long and strong, but a little shorter than the fourth of the dorsal. Pectorals pointed, elongate, not reaching to the vent.

Colour.—Silvery, the head, back, and upper part of sides thickly covered with small brown spots, of which there is one at the base of each scale; those on the occiput are smaller and closer. The scales on the lower part of the body have the base minutely punctulated with brown; the dorsal is spotted like the body, the other fins are immaculate.

Description.—This species resembles *P. nageb* and *punctulatum.* The greatest height of the body is below the fourth dorsal spine; it is nearly equal to the length of the head, and is contained thrice and two-thirds to four times in the total length. The upper profile descends very gently towards the nape and thence more obliquely, but with hardly any concavity, to the snout. The interorbital space is a little convex, and slightly larger than the eye. The snout is pointed, the under jaw rather overlapping the upper. The teeth form villiform bands, without canines. The præoperculum has the posterior limb emarginate, and the angle rounded but not produced backwards. The supra scapula is hardly serrated.

The origin of the dorsal is above the root of the upper pectoral ray, and its end behind that of the anal. The base of the soft portion is three-fourths that of the spinous; it is strongly notched; the spines are strong, broader on one side than on the other; the length of the first is three-sevenths that of the second; the second is hardly one-half that of the third; the third and fifth are nearly equal, and not much shorter than the fourth, which is about half the height of the body; the length of the penultimate spine is four-fifths of that before and after it; the last, which must be regarded as belonging to the soft portion, is three-fourths that of the first ray. The soft portion is lower than the spinous, the longest ray (fourth) being about equal to the seventh spine. The entire dorsal can be received into a scaly sheath.

The caudal is truncated; the lobes have rows of scales on the membranes between the rays to two-thirds of their length; these decrease in length towards the centre, which is scaly on the base only.

The anal spines are very strong, much broader on one side than on the other; the first is one-third the length of the third, and about equal to the fourth of the dorsal. This fin also can be received into a scaly sheath.

The pectoral is long and rather falciform, nearly as long as the head, but not reaching to the vent; the base is scaly to one-sixth of its length.

The ventrals are inserted slightly behind the posterior of the base of the pectorals. The first ray is about a third longer than the spine, excluding the extremity, which is slightly produced and filiform.

The scales are of moderate size; the tubules of the lateral line are bifurcate.

Coloration as above.

Length 12 inches.

Bagamoia, east coast of Africa. *A specimen from Port Natal exists in the British Museum.*

67.   Pristipoma operculare, sp. n.   Plate IV. fig. 1.   [69.]

D. 10 | $\frac{1}{14}$.   A. $\frac{3}{9-10}$.   L. lat. 56–60.   L. transv. 11–12/19.

Diagnosis.—Height of body equals the length of head, and is one-third of the total length without caudal. The snout is pointed, and so long that the distance of its extremity from the hind margin of the eye equals that between the end of operculum and centre of eye. Its length is nearly two-fifths of that of the head. Præorbital naked; the maxillary does not reach the vertical from the anterior nostril. Posterior margin of præoperculum serrated, but scarcely emarginate; the angle is rounded and armed with stronger teeth.

Dorsal fin deeply notched, the spines being long and strong; the fourth is longest, and equal to half the height of the body beneath it. Caudal emarginate. The second anal spine is slightly shorter than the fourth of the dorsal.

Colour.—Silvery, with numerous black spots on the back and sides. *A large black spot at the angle of the operculum*; a black spot at the base, in front of each dorsal spine and ray; anterior half of anal blackish.

Description.—The greatest depth of the body is below the fourth dorsal spine. The profile descends from the dorsal fin to the occiput in a slight curve, and is straight on the remainder of the head. The interorbital space is very convex, its width being more than the diameter of the eye. The snout is pointed, the upper jaw slightly overlapping the lower one. The posterior margin of the præoperculum is scarcely emarginate; the angle is rounded, but does not project backwards; the denticulations are stronger and wider apart at the angle. The suprascapula is distinct and entire.

The dorsal originates above the axil of the pectorals, and ends slightly behind the termination of the anal; the base of the soft portion is three-fifths of that of the spinous; it is deeply notched; the spines are strong, and broader on one side than on the other. The first is less than half the length of the second; the fourth is slightly longer than the third and fifth, and equal to half the height of the body below it; the penultimate spine is less than half the length of the fourth, and shorter than the second; the last must be regarded as belonging to the soft portion of the dorsal. The soft portion is much lower than the spinous, the longest ray (the fourth) being equal to the eighth spine. This fin, as also the anal, can be received into a scaly sheath. The caudal is emarginate; the lobes are scaly to three-fourths of their extent, and the centre to about one-half. The anal spines are strong; the first is very short; the second is the longest and strongest, and equal to the fourth of the dorsal. The pectorals are long, extending to the origin of the anal, and nearly as long as the head. The scales are moderate. The tubules of the lateral line are simple, and that line runs parallel to the upper profile of the fish.

Coloration as described above.

Length 11 inches.

Aden. *Two specimens also exist in the British Museum, received from Port Natal.*

68. Pristipoma maculatum. [651 n.]

Anthias maculatus, *Bloch*, t. 326. f. 2.
Caripe, *Russell*, pl. 124.
Pristipoma caripe, *Cuv. & Val.* v. p. 261 ; *Cantor, Mal. Fishes*, p. 75 ; *Rüpp. N. W. Fische*, p. 124.
—— maculatum, *Günth. Fish.* i. p. 293.

Bagamoia, east coast of Africa. *Red Sea. From Coromandel to New Guinea.*

E

69.   **Pristipoma stridens.**   [20.]

Sciæna stridens, *Forsk.* p. 50.
Pristipoma simmena, *Cuv. & Val.* v. p. 260.
—— stridens, *Rüpp. N. W. Fische*, p. 122, taf. 31. f. 1; *Günth. Fish.* i. p. 300.

Zanzibar. Aden. *Red Sea.*

DIAGRAMMA, *Cuv.*

70.   **Diagramma affine.**   [659.]

Pristipoma nigrum, *Cant. Mal. Fish*, p. 74 (an *Cuv. & Val.*??).
Diagramma affine, *Günth. Fish.* i. p. 319.
—— nigrum, *Day, Fishes of Malabar,* p. 23.

Zanzibar. *East-Indian archipelago. North-west coast of Australia.*

71.   **Diagramma griseum.** Plate IV. fig. 3, var. *b.*   [16, 330, 565.]

Diagramma griseum, *Cuv. & Val.* v. p. 306; *Günth. Fish.* i. p. 321.

$$\text{D. } \frac{12}{19\text{-}21}. \quad \text{A. } \frac{3}{7\text{-}8}. \quad \text{L. lat. } 74.$$

The specimen of this collection marked [16] is the variety described by Cuvier and Valenciennes. Nos. [330] and [565] are identical, and constitute a new variety, structurally the same as the previous, but differing widely as to coloration.

Var. *a* [16]. Uniform grey; fins darker. Length 13½ inches.
Zanzibar. Aden. *Malabar.*

Var. *b* [330, 565]. Colour grey above, white below, with four whitish curved cross bands; the first crosses the forehead and terminates at the angles of the operculum and præoperculum; the second proceeds from the second dorsal spine, in the direction of the root of the ventrals; the third runs parallel to the last from the seventh and eighth dorsal spines; and the fourth, also parallel, runs from the first anal ray to the posterior of anal. Fins blackish, immaculate. Length 4⅜ to 17 inches.
Zanzibar. Kiswarra Bay.

72.   **Diagramma pertusum.** Plate IV. fig. 2.   [532.]

Perca pertusa, *Thunberg, Nya Handl. Stockh.* 1793, xiv. p. 198, pl. 7. f. 1.
Diagramma thunbergii, *Cuv. & Val.* v. p. 308.
—— pertusum, *Günth. Fish.* i. p. 321.

$$\text{D. } \frac{10}{21\text{-}24}. \quad \text{A. } \frac{3}{6\text{-}7}. \quad \text{L. lat. } 100.$$

The upper profile, between the end of snout and commencement of dorsal, forms nearly a quadrant, the centre of which lies between the roots of the pectorals and ventrals ; the lower profile is nearly straight between the snout and anal. The diameter of the eye is contained three times and a half in the length of the head. The height of the body is contained three times and a half in the total length ; the length of the head is one-fourth of the same. The distance of end of dorsal from caudal is greater than the height of the tail below the former. Dorsal not notched, with moderately strong spines, of which the third is longest ; it is contained twice and a half in the length of the head ; the scaly sheath covers the basal third of the soft dorsal. The second anal spine is shorter but stronger than the third, which is contained twice and two-thirds in the length of the head. Caudal scarcely emarginate.

Colour uniform grey ; fins blackish. Length 10 inches.

Zanzibar. *Japanese seas.*

73.   **Diagramma centurio.**   [701.]

Diagramma centurio, *Cuv. & Val.* v. p. 308 ; *Günth. Fish.* i. p. 322.

$$D. \frac{10}{24-25}. \quad A. \frac{3}{7}. \quad L. \text{ lat. ca. } 105.$$

The height of the body is contained three times and a third, and the length of the head four times and a fifth in the total length. The diameter of the eye is contained four times in the length of the head, and once and a half in that of the snout. The upper maxillary does not reach the vertical from the front margin of orbit. Præoperculum with the posterior limb oblique, the angle rounded, the serrature strong, stronger and wider apart at the angle. Dorsal not notched ; the third spine is longest, and one-half the length of the head ; the soft portion is elevated posteriorly. Caudal emarginate.

Colour grey ; head, back, sides, upper part of tail, dorsal, and sometimes caudal with brownish-yellow spots, becoming blackish in a dried state.

Length 16 inches.

Seychelles.

74.   **Diagramma gaterina.**   [22.]

Sciæna gaterina, *Forsk.* p. 50.

Diagramma gaterina, *Cuv. & Val.* v. p. 301, pl. 125 ; *Rüpp. Atlas, Fische,* taf. 32. f. 1 ; *Günth. Fish.* i. p. 322.

Zanzibar. Aden. *Red Sea.*

75.   **Diagramma punctatum.**   [17, 84, 528.]

Diagramma punctatum, (*Ehrenb.*) *Cuv. & Val.* v. p. 302 ; *Rüpp. Atlas, Fische,* p. 126, t. 32. f. 2 ; *Quoy & Gaim. Voy. Astrolabe, Poiss.* pl. 12. f. 2, p. 699 ; *Günth. Fish.* i. p. 323.

—— cinerascens, *Cuv. & Val.* v. p. 307 (adult) ; *Rüpp. Atlas, Fische,* p. 127.

Zanzibar. Aden. *Red Sea. Seas of India, Java, and China.*

76.   **Diagramma pictum.**   [128.]

Perca picta, *Thunburg, Nya Handlingar Stockh.* 1793, xiii. p. 141, pl. 5.
Diagramma pictum, *Cuv. & Val.* v. p. 315; *Günth. Fish.* i. p. 327.

Zanzibar. Aden. *Indian seas. Penang. Java. Amboyna. Chinese and Japanese seas.*

77.   **Diagramma lessonii.**   [11, 531.]

Diagramma lessonii, *Cuv. & Val.* v. p. 313; *Less. Voy. Coq. Zool.* ii. p. 199, pl. 24 (bad) ; *Günth. Fish.* i. p. 329.

Var. *a.* With continuous blackish-brown longitudinal bands.
Var. *b.* With the longitudinal bands broken up into rows of round spots.
Zanzibar. Aden. *Waigiou. Amboyna.*

78.   **Diagramma blochii.**   [248, 636.]

Anthias diagramma, *Bl. tab.* 320.
Diagramma blochii, *Cuv. & Val.* v. p. 312 (?) ; *Günth. Fish.* i. p. 329.

$$D. \frac{9\text{-}10}{\overline{22}\text{-}\overline{23}}. \quad A. \frac{3}{7}.$$

The colour of the adult is dark grey, with darker longitudinal lines and series of spots; fins blackish, except pectorals, which are grey.

No. [636] is the adult of [248]. The younger specimen agrees perfectly with the *Anthias diagramma* of Bloch, although it is doubtful whether it is the same as the *D. blochii* of Cuvier and Valenciennes, which is only known from a figure taken at Trincomalee.

Zanzibar. Ceylon. Penang.

79.   **Diagramma cuvieri.**   [615, 618.]

Seba, iii. 27. 17.
Bodian cuvier, *Benn. Fishes of Ceylon,* no. 13.
Diagramma sebœ, *Bleek. Verh. Batav. Genootsch.* xxiii. *Sciæn.* p. 24; *Günth. Fish.* i. p. 331.

Zanzibar. Batavia. Banda Neira.

LOBOTES, *Cuv.*

80.   **Lobotes auctorum.**   [145.]

Holocentrum surinamensis, *Bl. t.* 243.
Lobotes erate, *Cuv. & Val.* v. p. 322 ; *Cantor, Mal. Fish.* p. 80 ; *Cuv. Règne An. Ill. Poiss.* pl. 31. f. 1.
—— auctorum, *Günth. Fish.* i. p. 338.

Mouth of Panganie River, east coast of Africa. *Indian Ocean and Archipelago. Atlantic shores of America. Caribbean Sea.*

SCOLOPSIS, *Cuv.**

81. **Scolopsis japonicus.** [4.]

Anthias japonicus, *Bloch*, t. 325. f. 2.
Kurite, *Russell*, pl. 106.
Scolopsis kurite, *Rüpp. Atlas, Fisch.* p. 9, t. 2. f. 3; *Cuv. & Val.* v. p. 331.
—— japonicus, *Günth. Fish.* i. p. 354.

Aden. *Red Sea. Coasts of India and China.*

82. **Scolopsis nototænia,** sp. n. Plate V. fig. 2. [113.]

D. $\frac{10}{9}$. A. $\frac{3}{7}$. L. lat. 44. L. transv. $\frac{5}{18}$.

Diagnosis.—The height of the body is contained three times and a half in the total length; length of the head thrice and three-fourths in the same. One infraorbital spine, the plate below it hardly serrated. Posterior limb of præoperculum emarginate, with the angle rounded and projecting backwards; denticulation conspicuous. The second and third anal spines nearly equal, the former is the stronger; caudal hardly notched.

Colour yellowish, *with a blackish band above the lateral line from below the fourth dorsal spine to the upper part of the tail, terminating at the commencement of the caudal.* A blue streak from angle of mouth across præorbital to eye.

Description.—This species somewhat resembles *S. ciliatus* in form, and *S. bimaculatus* in colour. The head is considerably smaller than in the former, and is covered with scales almost to the nostrils. The interorbital space is about three-fourths of the diameter of the eye, and less than the length of the snout. The cleft of the mouth is slightly oblique; the maxillary reaches the vertical from the anterior of the orbit. The præorbital is half as broad as the eye. The spine at the upper posterior angle is moderate, slightly fluted and pointed; the plate beneath it has a very few slight denticulations. The præoperculum has a narrow margin free from scales; the posterior limb descends obliquely forwards; the angle is rounded and projecting. The denticulations on the upper part of the posterior limb and those on the angle are strongest; the inferior limb is entire; it is considerably shorter than the posterior one. The operculum has a short prominent spine, and is covered with scales of rather small size. The suprascapula is tolerably distinct and minutely serrated.

The dorsal commences before the vertical from the base of pectorals, and terminates a little behind the anal. The base of the soft portion is about two-thirds as long as that of the spinous portion. The spines are moderately strong; the fourth to seventh are the longest. The former is contained twice and a third in the length of the head; the second

* Undetermined specimen [666].

is shorter than any behind it; the third is equal to the penultimate one. The posterior part of the soft is higher than the spinous portion. The distance between the dorsal and caudal is less than the height of the tail below the former. The caudal is very slightly forked, and is scaly at the base. The anal commences below the second, and extends to below the seventh dorsal ray; the two last spines are nearly equal in length, but the second is the stronger, they are about one-third of the length of the head; the first is less than half of the two succeeding ones; the soft portion is lower than the soft dorsal. Scales rather small.

Length 5½ inches.
Aden.

83.   **Scolopsis torquatus.**   [194.]

Scolopsis torquatus, *Cuv. & Val.* v. p. 335; *Günth. Fish.* i. p. 356; *Kner, Novara, Fisch.* p. 59.

Zanzibar. *Batavia. Molucca Sea. China.*

84.   **Scolopsis bimaculatus.**   [520.]

Scolopsis bimaculatus, *Rüpp. Atlas, Fisch.* p. 8, pl. 2. f. 2, and *N. W. Fisch.* p. 126; *Cuv. & Val.* v. p. 340; *Günth. Fish.* i. p. 357.

Zanzibar. *Red Sea. Ceylon. Chinese Sea.*

85.   **Scolopsis frenatus.**   [424.]

Scolopsis frenatus, *Cuv. & Val.* v. p. 343; *Günth. Fish.* i. p. 361.

Seychelles. *Mauritius.*

86.   **Scolopsis ghanam.**   [23, 404.]

Sciæna ghanam, *Forsk.* p. 50.
Scolopsis lineatus, *Rüpp. Atlas, Fisch.* p. 7, pl. 2. f. 1, and *N. W. Fisch.* p. 126.
—— ghanam, *Cuv. & Val.* v. p. 348; *Günth. Fish.* i. p. 362.

Zanzibar. Aden. *Red Sea.*

HETEROGNATHODON, *Bleek.*

87.   **Heterognathodon petersii.**

Heterognathodon petersii, *Steindachner, Sitzgsber. Akad. Wiss. Wien,* 1864, xlix. p. 203, t. 1. f. 2.

D. $\frac{10}{9}$.   A. $\frac{3}{7}$.   L. lat. 47.   L. trans. 3/11.

The length of the head is contained four times and a third in the total length, or three times and a third in the length without the caudal; the height of the body is about one-sixth of the total length; the diameter of the eye is one-third of the length of the head; and the width of the interorbital space is two-thirds of the diameter of the eye.

There are three or four rather strong canines on each side of the upper jaw.

The posterior margin of the præoperculum is finely and evenly serrated; the operculum terminates in a short spine.

The second spine of the anal is stronger but shorter than the third, which is shorter than the first ray. Both dorsal and anal fins can be received within a scaly sheath. The caudal is deeply forked; the lobes pointed, the upper being produced into a filament.

The upper part of the body is pale violet; from the posterior margin of the eye to the base of the caudal there is a very faint violet longitudinal band; below this the colour is bright yellow, the lower part of the belly being silvery. Ventrals yellow at the base. Dorsal with a narrow yellow marginal line. (*Steindachner.*)

Zanzibar.

SYNAGRIS, *Günth.*

88.　　　　　　　　　　**Synagris filamentosus.**　　　　　　　　[00, 183, 008.]

Cantharus filamentosus, *Rüpp. Atlas, Fisch.* p. 50, pl. 12. f. 3 (*not* Dentex filamentosus, *Cuv. & Val.*).
Dentex bipunctatus, (*Ehrenb.*) *Cuv. & Val.* vi. p. 217.
—— tambulus, *Cuv. & Val.* vi. pp. 219, 558; *Rüpp. N. W. Fisch.* p. 114.
Synagris filamentosus, *Günth. Fish.* i. p. 378.

The adult males only have the upper caudal lobe produced into a long filament.
Berbera. Aden. Zanzibar. *Red Sea. Pondicherry.*

PENTAPUS, *Cuv.*

89.　　　　　　　　　　**Pentapus curtus.**　　　　　　　　　　[633.]

Pentapus curtus, *Guichenot, in Maillard, Notes sur la Réunion, Poissons,* p. 5.
Zanzibar. *Réunion.*

CÆSIO, *Commers.*

90.　　　　　　　　　　**Cæsio lunaris.**　　　　　　　　　　[57.]

Cæsio lunaris, (*Ehrenb.*) *Cuv. & Val.* vi. p. 411; *Günth. Fish.* i. p. 390.
Zanzibar. Aden. *Red Sea. Batavia. New Ireland.*

91.   **Cæsio cærulaureus.**   [529.]

Cæsio cærulaureus, *Lacép.* iii. p. 86; *Cuv. & Val.* vi. p. 434; *Günth. Fish.* i. p. 392; *Kner, Novara, Fisch.* p. 65.

$$D. \frac{10}{13}. \quad A. \frac{3}{11-15}. \quad L. \text{ lat. } 55.$$

Zanzibar.   *Mauritius.*   *Red Sea.*   *Ceylon.*

92.   **Cæsio striatus.**   [223, 702.]

Cæsio striatus, *Rüpp. Atlas, Fische,* p. 131, pl. 34. f. 1; *Günth. Fish.* i. p. 392.

Zanzibar.   Seychelles.   *Red Sea.*

Family SQUAMIPINNES.

CHÆTODON, *Artedi.*

93.   **Chætodon strigangulus.**   [244.]

Chætodon strigangulus, (*Solander*) *Gm.* p. 1269; *Cuv. & Val.* vii. p. 42, pl. 172; *Beechey, Voy. Zool.* p. 60, pl. 17. f. 2; *Günth. Fish.* ii. p. 4.
—— triangularis, *Rüpp. Atlas, Fische,* p. 42, pl. 9. f. 1.

Zanzibar.   *From Red Sea to Polynesia.*

94.   **Chætodon setifer.**   [170.]

Chætodon setifer, *Bl.* t. 426. f. 1; *Cuv. & Val.* vii. p. 76; *Less. Voy. Coq. Zool.* ii. p. 175, *Poiss.* pl. 29. f. 2; *Günth. Fish.* ii. p. 6.

Chætodon auriga, vâr., *Rüpp. N. W. Fische,* p. 20.

Zanzibar.   Seychelles.   *From Red Sea to Polynesia.*

95.   **Chætodon unimaculatus.**   [524.]

Chætodon unimaculatus, *Bloch,* t. 201. f. 1; *Cuv. & Val.* vii. p. 72; *Günth. Fish.* ii. p. 11.

$$D. \frac{13}{21-23}. \quad A. \frac{3}{18-21}.$$

This specimen differs from those in the British Museum in having the ocular band much narrower than the eye; it also terminates on the opercles, without being continued round the chest. Above the orbit the band is one-half, and below it less than one-third the diameter of the eye.

Zanzibar.   *Molucca Sea.*   *Polynesia.*   *Micronesia.*

96. **Chætodon bennetti.** [561.]

Chætodon bennetti, *Cuv. & Val.* vii. p. 8 ł ; *Günth. Fish.* ii. p. 12.

This specimen differs slightly from that in the British Museum in the number of dorsal spines and rays, which are $\frac{13}{16}$. The caudal has a broad white margin ; the other fins are uniform deep yellow.

Zanzibar. *Molucca Sea.*

97. **Chætodon zanzibarensis,** sp. n. Plate VI. fig. 1. [181.]

D. $\frac{14}{17}$. A. $\frac{3}{16}$. L. lat. 48.

Diagnosis.—Snout very slightly produced, and rather shorter than the diameter of the eye ; præoperculum slightly denticulated. Dorsal and anal fins rounded posteriorly. The black ocular band is half the diameter of the eye, and is continued over the chest. A large black blotch on the side below the four posterior spines and the three first rays of the dorsal ; two-thirds of this blotch is below the lateral line. Caudal with a broad white margin ; the remainder of the body and fins yellow, with narrow darker longitudinal lines following the series of scales.

Description of the specimen.—Body oval, its greatest height being half of the total length. The upper profile descends abruptly from the origin of the dorsal ; the lower profile is very similar to the upper. The snout is short, obtuse, rather less than the diameter of the eye, or than the interorbital space. The angle of the præoperculum is nearly a right one, rounded and slightly denticulated. The spines of the dorsal are moderately strong ; the fifth and sixth are the longest ; thence they decrease slightly in length, the last being equal to the fourth. The soft portion is lower than the spinous, and is rounded posteriorly. The caudal is convex ; the soft portion of the anal corresponds to that of the dorsal. The second anal spine is longest and strongest, but shorter than the longest of the dorsal. The pectorals and ventrals do not reach as far as the vent. Scales moderate ; one of the largest covers half the eye.—Length $5\frac{1}{2}$ inches.

Zanzibar.

98. **Chætodon falcula.** [619.]

Chætodon falcula, *Bloch,* ix. p. 102, t. 426. f. 2 ; *Cuv. & Val.* vii. p. 41 ; *Günth. Fish.* ii. p. 17.

Zanzibar. *Sea of Batoe.*

99. **Chætodon kleinii.** [139.]

Klein, Miss. iv. tab. 10. f. 255.

Chætodon kleinii, *Bloch,* tab. 218. f. 2 ; *Günth. Fish.* ii. p. 22.

—— melastomus, *Bl. Schn.* p. 224.

—— virescens, *Cuv. & Val.* vii. p. 30.

—— flavescens, *Benn. Proc. Comm. Zool. Soc.* i. p. 61.

Zanzibar. *Indian Ocean and Archipelago.*

F

100. **Chætodon trifasciatus.** [230.]

Chætodon trifasciatus, *Mungo Park, Trans. Linn. Soc.* iii. p. 34; *Lacép.* x. p. 498.
—— vittatus, *Bl. Schn.* p. 227; *Cuv. & Val.* vii. p. 34; *Günth. Fish.* i. p. 23; *Kner, Novara, Fische,*
p. 100 (*not C. austriacus, Rüpp.*).

Zanzibar. *From the Red Sea to Polynesia.*

101. **Chætodon melanopterus.** [5.]

Chætodon melapterus, *Guichenot in Maillard, Notes sur la Réunion, Annexe C. Poissons,* p. 6.

$$\text{D. } \frac{13}{21}. \quad \text{A. } \frac{3}{28}.$$

Scales large; snout conical, a little longer than the diameter of the eye; præoperculum hardly serrated; dorsal and anal rounded posteriorly.

Ground-colour of body brilliant yellow, with darker longitudinal stripes following the series of scales. Guichenot, in his description of a specimen found at Réunion, says, "et semé partout de très-petits points noirs, qui font paraître le corps comme sablé;" in the specimens from Aden, however, there is no appearance of such punctulation. The extremity of the snout is black, as is the ocular band and another parallel to it descending from the first dorsal spines to the angle of the præoperculum; the ocular band is narrower than the orbit and extends round the throat. The vertical fins are entirely black, except the margins of the spinous dorsal and anal, which are yellowish, and the posterior quarter of the caudal, which is white.

Aden. *Réunion.*

102. **Chætodon lunula.** [538.]

Pomacentrus lunula, *Lacép.* iv. pp. 507, 510, 513.
Chætodon lunula, *Cuv. & Val.* vii. p. 59, pl. 173; *Günth. Fish.* ii. p. 25.

Zanzibar. *Indian Ocean and Archipelago. Polynesia.*

103. **Chætodon vagabundus.** [7.]

Chætodon vagabundus, *Linn. Syst. Nat.* i. p. 465; *Bl.* t. 204. f. 2; *Cuv. & Val.* vii. p. 50; *Günth. Fish.* ii. p. 25.

Aden. *Red Sea to Polynesia.*

104. **Chætodon guttatissimus.** [515.]

Chætodon guttatissimus, *Bennett, Proc. Comm. Zool. Soc.* ii. p. 183; *Günth. Fish.* ii. p. 26.

Zanzibar. *Ceylon.*

105. **Chætodon dorsalis.** [217.]

Chætodon dorsalis, (*Reinw.*) *Cuv. & Val.* vii. p. 70; *Rüpp. Atlas, Fische,* p. 41, t. 9. f. 2; *Günth. Fish.* ii. p. 28.

Zanzibar. *From Red Sea to Polynesia.*

106. **Chætodon ocellicauda.** [587.]

Chætodon ocellicauda, *Cuv. & Val.* vii. p. 69; *Günth. Fish.* ii. p. 29.

Zanzibar. *Indian Ocean.*

107. **Chætodon melanopoma,** sp. n. Plate VI. fig. 2. [129.

D. $\frac{12}{25}$. A. $\frac{3}{21}$. L. lat. 25.

Diagnosis.—Scales large. Snout slightly produced and a little longer than the diameter of the eye; præoperculum very indistinctly serrated; dorsal and anal fins with the posterior margins rounded. The lateral line ascends from the suprascapula to the base of the fourth dorsal ray, runs along the base of the soft dorsal, and is lost towards the end of that fin; the ocular band assumes the appearance of a dark blotch, extending in breadth nearly to angle of operculum, and in length to extremity of suboperculum.

Colour.—Yellow, with darker lines, nearly vertical, following the series of scales. The soft portion of the dorsal, the anal, and the caudal have light margins and black intramarginal lines.

Description of the Specimen.—The body is rather elevated, its greatest height being contained once and two-thirds in the total length. The upper profile descends abruptly, there being a long concavity in front of the eye; the inferior profile is lower and less convex. The angle of the præoperculum is nearly a right one, rounded and very indistinctly serrated. The spines of the dorsal are rather strong, increasing in length posteriorly, the last being the longest; the soft is slightly more elevated than the spinous portion, and is rounded posteriorly. The caudal is slightly convex. The soft portion of the anal is similar to that of the dorsal. The second anal spine is rather shorter than the third, which is about equal to the fourth of the dorsal. The scales are large, one of the largest being nearly as high as the diameter of the eye. The ventrals reach as far as the anal, and the first rays are slightly produced.

Aden.

108. **Chætodon lineolatus.** [208.]

Chætodon lineolatus, *Cuv. & Val.* vii. p. 40; *Günth. Fish.* ii. p. 30.
—— lunatus, *Cuv. & Val.* vii. p. 57; *Rüpp. N. W. Fische,* p. 30, t. 9. f. 3.

Zanzibar. *Red Sea. Sea of Floris.*

109. **Chætodon leucopleura,** sp. n. Plate VI. fig. 3. [511.]

D. $\frac{12}{21-24}$. A. $\frac{3}{26}$. L. lat. 27. L. transv. 5/13.

Diagnosis.—Scales large. Snout produced and pointed; its length is once and a half the diameter of the eye. Præoperculum entire. The soft vertical fins with an obtuse angle behind. Caudal slightly emarginate, the upper lobe a little produced. Dorsal

F 2

spines low, the fifth longest, and equal to the extent of the snout. Anal spines strong, the second stronger and scarcely longer than the third.

Colour.—Ocular band rather narrower than the orbit, brown above and below and yellow on the cheek ; it extends from the origin of the dorsal, round the throat. Back and belly brownish, shaded off into white or yellowish on the middle of the sides. Five or six yellow longitudinal stripes on the belly, the highest proceeding from the axil of the pectoral. Tail yellow; caudal yellow, with a broad white or grey margin; vertical fins yellow, with a grey intramarginal line on the soft portion. Posterior margin of operculum yellow ; ventrals yellow; pectorals white. Muzzle blackish, with several yellow lines across the interorbital space, two yellow circular marks on the neck.

Description of the Specimen.—The body is rather oval, the greatest height being below the sixth dorsal spine, where it is contained twice and a quarter in the total length. The upper profile descends rather obliquely from the origin of the dorsal, and is slightly concave in front of the orbit; the lower profile also is concave below the angle of the mouth. The snout is somewhat produced, pointed, and its length is once and a half the diameter of the eye. The angle of the præoperculum is very obtuse, the limbs being entire. The spines of the dorsal are short and strong, the fifth, sixth, and seventh being longest and nearly equal, and about the same length as the snout; the succeeding spines become shorter, the last being slightly shorter than the fourth; the soft portion is scarcely elevated posteriorly and is obtusely pointed. The caudal is slightly emarginate, the upper lobe being somewhat produced. The soft portion of the anal resembles that of the dorsal ; the middle spine is stronger, but hardly shorter than the third, which is longer than the longest of the dorsal; the first of the anal equals the third of the dorsal. The pectorals reach the vertical from the vent, the ventrals do not. The scales are large, especially those on the middle of the sides; they are rather irregularly arranged, and smaller scales are intercalated between the larger ones. The lateral line ascends to below the tenth dorsal spine, and runs backwards not far below the base of the dorsal, towards the end of that fin, having entirely disappeared on the free portion of the tail. Length 7½ inches.

Zanzibar.

110.   **Chætodon larvatus.**   [6.]

Chætodon larvatus, (*Ehrenb.*) *Cuv. & Val.* vii. p. 45; *Rüpp. N. W. Fische,* p. 28; *Günth. Fish.* ii. p. 31.

Aden.   *Red Sea.*

111.   **Chætodon xanthocephalus.**   [625.]

Chætodon xanthocephalus, *Benn. Proc. Comm. Zool. Soc.* ii. p. 182; *Günth. Fish.* ii. p. 33.

Zanzibar.   *Ceylon.*

HENIOCHUS, *Cuv. & Val.*

112.   **Heniochus macrolepidotus.**   [8.]

Seba, iii. 25. 8.
Chætodon macrolepidotus, *Artedi, Species,* p. 94; *Linn. Syst. Nat.* i. p. 464.
Heniochus macrolepidotus, *Cuv. & Val.* vii. p. 93; *Schleg. Faun. Japon. Poiss.* p. 82, pl. 44. f. 1
(young); *Günth. Fish.* ii. p. 39.
Diphreutes macrolepidotus, *Cant. Mal. Fish.* p. 159.

Aden. Zanzibar. Seychelles. *Through all the Indian seas to Australia.*

HOLACANTHUS, *Lacép.*

113.   **Holacanthus asfur.**   [1.]

Chætodon asfur, *Forsk.* p. 61.
Holacanthus asfur, *Rüpp. Atlas, Fische,* p. 132, t. 34. f. 2; *Cuv. & Val.* vii. p. 174; *Günth. Fish.* ii. p. 45.

Zanzibar. Aden. *Red Sea.*

114.   **Holacanthus diacanthus.**   [243.]

Chætodon diacanthus, *Boddaert, Epist. ad Gaub. de Chæt. diac.* 1772.
—— fasciatus, *Bloch,* t. 195.
Holacanthus dux, *Lacép.* iv. p. 534; *Cuv. & Val.* vii. p. 184; *Rüpp. N. W. Fische,* p. 37.
—— diacanthus, *Günth. Fish.* ii. p. 48.

Zanzibar. *Indian Ocean and Archipelago.*

115.   **Holacanthus multispinis,** sp. n. Plate VI. fig. 4.   [491.]

$$D. \frac{14}{16}. \quad A. \frac{3}{17}. \quad L. \ lat. \ 47.$$

Præopercular spine flat, smooth, reaching to base of pectorals; a smaller spine in front of it. The most remarkable characteristic of this fish is that there are from two to four spines on the interoperculum, of which the posterior is longest, and longer than the additional spine of the præoperculum. Suborbital strongly denticulated. Soft dorsal and anal subrhomboidal, not elevated. Anal spines very long; the first is equal to the fourth of the dorsal; the third is equal to the last; the second is intermediate between them. The height of the body below the origin of the dorsal is one-third of the total length; the length of the head to the angle of operculum is contained four times and two-thirds in the same. The snout is very obtuse.

Colour in spirits and dried examples uniform dark silky brown or black. In life deep brown, darker anteriorly, with numerous black wavy bands, some of which are interrupted. A blue longitudinal band parallel to anal. On the shoulder there is a black blotch with a bluish border. Dorsal with a black basal band and two similar parallel ones on the soft portion. Anal with three dark parallel bands and a brilliant blue margin. Ventral with blue membrane between the spine and first ray. Pectorals black. The markings disappear rapidly after death.

Length 4½ inches.

Zanzibar.

116.   **Holacanthus trimaculatus.**   [473.]

Holacanthus trimaculatus, (*Lacép.*) *Cuv. & Val.* vii. p. 196, pl. 182 ; *Günth. Fish.* ii. p. 50.

Zanzibar. *Molucca Sea.*

117.   **Holacanthus imperator.**   [205.]

Chætodon imperator, *Bloch,* iii. p. 51, t. 194.

Holacanthus imperator, *Lacép.* iv. pp. 527, 534, pl. 12. f. 3 ; *Cuv. & Val.* vii. p. 180 ; *Günth. Fish.* ii. p. 52.

Zanzibar. *Indian Ocean and Archipelago.*

118.   **Holacanthus semicirculatus.**   [599.]

Holacanthus semicirculatus, *Cuv. & Val.* vii. p. 191, pl. 183; *Less. Voy. Cogn. Zool. Poiss.* p. 173, pl. 30. f. 3 ; *Günth. Fish.* ii. p. 53.

Zanzibar. *East-Indian archipelago. Polynesia.*

119.   **Holacanthus alternans.**   [196. 216.]

Holacanthus alternans, *Cuv. & Val.* vii. p. 193 ; *Günth. Fish.* ii. p. 53.

In old specimens [196] all traces of cross bands disappear, but the darker spots on the head and body remain.

The affinities between *H. semicirculatus* and *H. alternans* are of such a nature as to call for a renewed examination of fresh specimens to settle the question whether the differences between them are not dependent on sexual development or age. Specimens with cross bands are 6 inches long, those without bands 10¾ inches.

Zanzibar. Aden. *Madagascar.*

120.   **Holacanthus chrysurus.**   [52, 348.]

Holacanthus chrysurus, *Cuv. & Val.* vii. 188 ; *Günth. Fish.* ii. p. 51.

$$D. \frac{12\text{-}13}{20\text{-}19}. \quad A. \frac{3}{19\text{-}20}.$$

Immature specimens [52] have a broad yellow band from the last dorsal spine and three first rays across the body nearly to the anal.

Aden. Zanzibar.

SCATOPHAGUS, *Cuv. & Val.*

121. **Scatophagus tetracanthus.** [849.]

Chætodon tetracanthus, *Lacép.* iv. p. 727, iii. pl. 25. f. 2.
Scatophagus fasciatus, *Cuv. & Val.* vii. p. 144.
—— tetracanthus, *Günth. Fish.* ii. p. 60.

$$\text{D. } 10 \mid \tfrac{1}{16}. \quad \text{A. } \tfrac{4}{14\text{-}15}.$$

The width between the eyes is contained twice and a third in the length of the head. The head is two-sevenths of the total length without caudal ; the greatest height of the body is above the first anal spine ; the fourth dorsal spine is about one-third of this height ; the præorbital is not denticulated.

Colour reddish olive, with six brown cross bands : the first from the nape, over the eye ; the second from before the first dorsal spine, over the operculum ; the third below the fourth to sixth dorsal spines ; the fourth below the last three dorsal spines ; the fifth below the middle of the soft dorsal ; and the sixth across the tail.

Length 6 inches.
Mouth of Pangani River, east coast of Africa.

DREPANE, *Cuv. & Val.*

122. **Drepane punctata.** [111.]

Chætodon punctatus, *L. Gm.* p. 1243.
Russell, pls. 79, 80, 81.
Drepane punctata, *Cuv. & Val.* vii. p. 132, pl. 179 ; *Günth. Fish.* ii. p. 62 ; *Kner, Novara, Fisch.* p. 107.

East coast of Africa. Red Sea, *through Indian Ocean to north-west coast of Australia.*

Family NANDIDÆ.

PLESIOPS, *Cuv.*

123. **Plesiops nigricans.** [843.]

Pharopteryx nigricans, *Rüpp. Atlas, Fische,* p. 15, t. 4. f. 2, and *N. W. Fische,* p. 5.
Plesiops nigricans, *Günth. Fish.* iii. p. 363.

A single specimen, length 3⅔ inches, was found in fresh water at Johanna. *Red Sea.*

Family MULLIDÆ.

It appears to us that the divisions of this family, established by Bleeker, are not of generic value; we therefore propose to reunite them all under the genus *Mullus* of Linnæus.

MULLUS, *L.* *

124.   **Mullus vittatus.**   [60, 336.]

Mullus vittatus, *Forsk.* p. 31; *Russell*, pl. 158.
Upeneus vittatus *et* bivittatus, *Cuv. & Val.* iii. p. 448, vii. p. 520; *Rüpp. N. W. Fische*, p. 101.
—— bitæniatus, *Benn. Proc. Comm. Zool. Soc.* 1830–31, p. 59.
Upeneoides vittatus, *Günth. Fish.* i. p. 397.

Aden. Zanzibar. Mayotta. *Red Sea, and nearly all Indian seas.*

125.   **Mullus tragula.**   [19, 272, 675, 690.]

Upeneus tragula, *Richardson, Ichth. China*, p. 220.
Upeneoides variegatus, *Bleek. Verhand. Bat. Genootsch.* xxii. 1849, *Perc.* p. 64, and *Act. Soc. Indo-Neerl.* ii. *Amb.* p. 48.
Upeneoides tragula, *Günth. Fish.* i. p. 398.

Aden. Zanzibar. *East-Indian archipelago. Chinese seas.*

126.   **Mullus barberinus.**   [564.]

Mullus barberinus, *Lacép.* p. 406, pl. 13. f. 3.
Upeneus barberinus, *Cuv. & Val.* iii. p 462; *Rüpp. N. W. Fische*, p. 101; *Günth. Fish.* i. p. 405.

Zanzibar. *Red Sea, East-Indian seas.*

127.   **Mullus micronemus.**   [601, 161.]

Mullus micronemus, *Lacép.* iii. p. 405, pl. 13. f. 2.
—— auriflamma, *Lacép.* iii. p. 400 (not *Forsk.*).
Upeneus lateristriga, *Cuv. & Val.* iii. p. 463; *Rüpp. N. W. Fische*, p. 101.
—— micronemus, *Günth. Fish.* i. p. 405.

Zanzibar. Johanna. *Red Sea.*

128.   **Mullus pleurostigma.**   [557.]

Upeneus pleurostigma, *Benn. Proc. Comm. Zool. Soc.* i. p. 59.
Upeneus brandesii, *Bleek. Nat. Tydsch. Nederl. Ind.* 1851, *Banda*, i. p. 236; *Günth. Fish.* i. p. 407.

Zanzibar. *Mauritius. Banda Neira.*

* Undetermined specimen [563].

129.   **Mullus malabaricus.**   [25.]

Upeneus malabaricus, *Cuv. & Val.* iii. p. 467; *Günth. Fish.* i. p. 407.

Aden.   Zanzibar.   *Malabar.*   *Philippines.*

130.   **Mullus oxycephalus.**   [540.]

Upeneus oxycephalus, *Bleek. Act. Soc. Sc. Indo-Nederl.* i. *Manado en Makass.* p. 45; *Günth. Fish.* i. p. 409.

Zanzibar.   *Mauritius.*   *Sea of Manado.*

131.   **Mullus luteus.**   [51, 218, 367, 850.]

Upeneus luteus, *Cuv. & Val.* iii. p. 521.

It is doubtful whether all the specimens here mentioned belong to the same species; but, as they are mostly stuffed, it is impossible to separate them with certainty.

Zanzibar. Aden. *Indian Ocean.*

132.   **Mullus dispilurus,** sp. n.   Plate V. fig. 3.   [335, 477.]

D. 8 | 9.   A. $\frac{1}{7}$.   L. lat. 29.   L. transv. 2/7.

No teeth on the palate; those of the jaws in a single series. The height of the body is contained four times, and the length of the head four times and a quarter in the total length. The snout is of moderate length, the anterior margin of the eye occupying the middle of the head; the upper profile descends in a regular and gentle curve. The interorbital space is very convex, much broader than the diameter of the eye, and equal to half the length of the snout. The barbels hardly reach the posterior limb of the præoperculum. The tubules of the lateral line are moderately branched. The height of the spinous dorsal is nearly half that of the body.

A black patch occupies the back of the tail, extending from the lateral line of one side to that of the other; between this and the base of the dorsal is a bright patch, also extending between the lateral lines.

Length 4½ to 11 inches.

Zanzibar. Island of Pemba.

133.   **Mullus pleurotænia,** sp. n.   Plate V. fig. 4.   [519.]

D. 8 | 9.   A. 7.   L. lat. 29.   L. transv. 2½/7.

No teeth on the palate; those of the jaws in a single series. The height of the body equals the length of the head, and is one-fourth of the total length. Snout of moderate length, the upper profile descending in a regular gentle curve. Interspace between the eyes convex, and contained once and a half in the length of the snout. The barbels reach slightly beyond the vertical from the posterior margin of the præoperculum. The spinous dorsal is nearly two-thirds of the height of the body.

Colour in life.—Body marbled with greenish and rosy, each scale with a green margin.

Two shining longitudinal bands, one from the posterior of upper margin of orbit, above the lateral line, towards the middle of the soft dorsal; and the second and broader from the upper lip, below the eye and lateral line, towards the middle of the back. A bright spot on the back of the tail, immediately behind the second dorsal. First dorsal rosy, marbled with lighter and with brown. The second dorsal with four, and the anal with three, longitudinal series of rosy spots.—Length 7 inches.

Zanzibar.

Family SPARIDÆ.

TRIPTERODON, g. n.

We place here, in the group *Cantharina*, a fish very similar in habit to *Ephippus orbis*, but distinguished from that fish chiefly by its teeth. We regret that we have had but a single stuffed specimen under observation; fortunately, however, that was an adult, in remarkably fine condition.

Generic Characters.—Body much compressed and elevated; snout short; upper profile parabolic. Dorsal deeply emarginate between the spinous and soft portions, the former not scaly, but folding into a groove; soft portions of vertical fins covered with scales. Pectoral short, much shorter than the ventrals; anal spines three. Scales sparoid, granulated. *Large moveable tricuspid teeth in several series in both jaws*, none on the vomer or palatines. Bones of the jaws very short, cleft of mouth narrow; præorbital elevated, much higher than the orbit. Gill-opening of moderate width, separated from its fellow by a broad isthmus; gill-membranes not extending across the throat.

134.   **Tripterodon orbis**, sp. n.   Plate VII. fig. 1.   [626.]

D. 8 $\frac{1}{20}$. A. $\frac{3}{16}$. L. lat. 52. L. transv. 11/28.

The body is compressed and greatly elevated, the upper profile being most strongly curved: it descends very abruptly from the origin of the dorsal to the snout, forming a rather prominent protuberance between the eyes. The greatest height of the body is below the fifth dorsal spine, where it is one-half of the total length; the length of the head is contained four times and two-thirds in the same. The eye is situated high up in the head; its diameter is about half the length of the snout, and is contained thrice and a half in the length of the head. The cleft of the mouth is small, the upper maxillary reaching to a vertical from the posterior nostril. The posterior processes of the intermaxillary are rather short, the upper jaw being but slightly protractile. The præorbital is much higher than the orbit, and, like the upper part of the skull, it is naked. The interorbital space is concave. There are five series of scales on the cheek. The posterior limb of the præoperculum is naked, and the whole of the operculum is covered with scales. The præoperculum is obtusely denticulated, and the operculum terminates in a conspicuous point posteriorly. The suprascapula is concealed by the scales. In the upper jaw there are four, and in the lower jaw three, series of large, broad, flat, moveable, tricuspid teeth, there being about twenty in the outer series of the upper jaw.

The dorsal spines are very broad and flat, but rather feeble; the first and second are minute; the third, fourth, and fifth are produced into long filaments, which are about twice the length of the head; the sixth is less than the diameter of the eye; the seventh and eighth decrease in length; the ninth, which must be regarded as belonging to the soft portion, is shorter than the sixth; all these spines are inserted in, and can be received within, a deep scaly sheath: the soft portion is scaly, the length of its base is greater than that of the spinous portion; it is elevated anteriorly, the longest rays being not quite as long as the head; the upper margin is rounded; the first ray is half the length of the second; the next six are nearly equal; thence they gradually decrease in length; the last is the shortest, and is about one-third the length of the second. Caudal emarginate. Anal elevated anteriorly, its lower edge is emarginate: the spines are short, strong, and rather remote from each other; the second and third are nearly equal in length: the soft portion is covered with scales; the third ray is longer than the longest of the dorsal. Pectoral obtusely pointed, and much shorter than the ventral. Ventral long, the first ray being produced into a filament which reaches to the anal.

The scales are of moderate size, sparoid, higher than broad; at the axil of the ventral they are modified into a lanceolate flap.

Colour silvery grey, with about nine indistinct darker cross bands: the first (through the eye) and the second (from the top of the interparietal crest, passing over the superior angle of operculum and root of pectorals) are more distinct than the others.

Zanzibar.

Length 11¼ inches.

SARGUS, Klein.

135. <div align="center">**Sargus rondeletii.**</div> [166.]

Sparus sargus, *L. Gm.* p. 1270; *Bl.* t. 264.
Sargus raucus, *Geoff. Descr. Egypt. Poiss.* pl. 18. f. 1.
—— rondeletii, *Cuv. & Val.* vi. p. 14, pl. 14; *Günth. Fish.* i. p. 440.

This specimen was obtained at Maculla, on the south coast of Arabia, and was the only one observed by Colonel Playfair either there or elsewhere on the Arabian and African coasts. We have compared it with specimens both of *S. rondeletii* and *S. capensis*; and it is singular that it agrees better with the Mediterranean than with the African species.

South Arabia. *Mediterranean. Madeira. Canary Islands.*

LETHRINUS, *Cuv.*

136.   **Lethrinus latifrons.**   [204.]

Lethrinus latifrons, *Rüpp. N. W. Fische*, p. 118, t. 28. f. 4; *Günth. Fish.* i. p. 458.

Zanzibar.   *Red Sea.   East-Indian seas.*

137.   **Lethrinus longirostris,** sp. n.   Plate VII. fig. 2.   [364.]

D. $\frac{10}{9}$.   A. $\frac{3}{8}$.   L. lat. 48.   L. transv. 6/17.

Diagnosis.—The height of the body is contained four, and the length of the head three times and a half in the total length. *Snout very much elongated and pointed, the posterior nostril occupying the middle of the length of the head.* The diameter of the eye is contained five times and a half in the length of the head; and the distance between the eyes is about equal to the diameter of the orbit. No true molars, canines moderate, teeth of lateral series slightly compressed and pointed. Caudal emarginate. The third and fourth dorsal spines longest, and longer than the longest ray; ventrals and pectorals reach as far as vent. Violet, with darker markings about the head and snout.

Description of the specimen.—This species is low, elongated, and with an unusually long snout. The greatest height of the body is above the base of the ventrals, where it is one-fourth of the total length. The upper profile of the fish, from the origin of the dorsal to the occiput, is nearly horizontal; thence it descends in a straight line, forming an angle of forty degrees with the axis of the fish.

The length of the head is contained thrice and a half in the total length. The length of the intermaxillary process equals the distance from the anterior margin of the orbit to the extremity of the operculum. The maxillary bone does not nearly reach the vertical from the anterior nostril. The eye is situated far back in the head; its diameter is one-third of the length of the snout, and it is contained five times and a half in that of the head. The lips are thick and fleshy. The operculum is scaly and terminates in a very distinct point, with a rather deep curve above it. The suboperculum also is scaly, the præ- and interoperculum are naked. The scapular and humeral bones are distinctly visible and entire.

The dorsal spines are slender, the third and fourth being longest; the former is contained thrice and a half in the length of the head, and is twice the length of the first. The longest of the soft rays of the dorsal fin (the fifth) is equal in length to the third spine.

The distance between the end of the dorsal and the root of the caudal is about equal in length to the base of the soft dorsal.

The caudal is forked, the lobes being pointed.

The anal spines are rather stouter than those of the dorsal; the first is half the length of the third, and the second is intermediate in length between them; the third is equal to the last of the dorsal, and is about one-fifth of the length of the head.

The third pectoral ray is the longest, and reaches to the vent; it is contained five times and two-thirds in the total length.

The ventral spine is long and slender, and longer than the longest of the dorsal.

There are four canines in each jaw, of which the exterior ones are the longest; there are no true molars; the teeth of the lateral series are slightly compressed and pointed, the three or four in the middle being the largest.

The tubules of the lateral line are distinct and simple.—Length 11 inches.

Zanzibar.

138. **Lethrinus harak.** [613.]

Sciæna harak, *Forsk.* p. 52.

Lethrinus harak, *Rüpp. N. IV. Fische,* p. 116, t. 29. f. 3 ; *Günth. Fish.* i. p. 458.

Zanzibar. *Red Sea. East-Indian seas.*

139. **Lethrinus ramak.** [663.]

Sciæna ramak, *Forsk.* p. 52.

Lethrinus ramak, *Rüpp. N. IV. Fische,* p. 117, t. 18. f. 3 ; *Günth. Fish.* i. p. 459.

Zanzibar. *Red Sea.*

140. **Lethrinus nebulosus.** [2, 31.]

Sciæna nebulosa, *Forsk.* p. 52.

Lethrinus nebulosus, *Cuv. & Val.* vi. p. 284; *Rüpp. N. W. Fische,* p. 118; *Günth. Fish.* i. p. 460.

—— centurio, *Cuv. & Val.* vi. p. 301 ; *Peters, Wiegm. Arch.* 1855, p. 243.

—— esculentus, *Cuv. & Val.* pl. 158.

Aden. Zanzibar. Seychelles. *Red Sea. Mozambique.*

PAGRUS, *Cuv.*

141. **Pagrus spinifer.** [362.]

Sparus spinifer, *Forsk.* p. 32.

Russell, ii. p. 1, pl. 101.

Pagrus spinifer, *Cuv. & Val.* vi. p. 156; *Rüpp. N. W. Fische,* p. 114; *Günth. Fish.* i. p. 472.

Zanzibar. *Red Sea. East-Indian seas.*

CHRYSOPHRYS, *Cuv.*

142. **Chrysophrys sarba.** [543.]

Russell, pl. 91.

Sparus sarba, *Forsk.* p. 31.

Chrysophrys sarba, *Cuv. & Val.* vi. p. 102 ; *Rüpp. N. W. Fische,* p. 110, pl. 28. f. 1 ; *Günth. Fish.* i. p. 488.

Zanzibar. *Mauritius. Red Sea to Molucca Sea.*

143.   **Chrysophrys bifasciata.**   [349.]

Chætodon bifasciatus, *Forsk.* p. 64.

Chrysophrys bifasciata, *Cuv. & Val.* vi. p. 118; *Rüpp. N. W. Fische,* p. 112; *Günth. Fish.* i. p. 488.

Zanzibar. *Red Sea. East-Indian seas.*

PIMELEPTERUS, *Lacép.*

144.   **Pimelepterus lembus.**   [611.]

Pimelepterus lembus, *Cuv. & Val.* vii. p. 269; *Günth. Fish.* i. p. 498.

Zanzibar. *Batavia. Vanicolo.*

145.   **Pimelepterus tahmel.**   [274.]

Sciæna tahmel, *Forsk.* p. 53.

Pimelepterus altipinnis, *Cuv. & Val.* vii. p. 270.

—— tahmel, *Rüpp. N. W. Fische,* p. 35, t. 10. f. 4; *Günth. Fish.* i. p. 499.

Zanzibar. *From Red Sea to New Guinea.*

Family CIRRHITIDÆ.

CIRRHITES, *Commers.**

146.   **Cirrhites forsteri.**   [571.]

Seba, iii. 27, 12.

Grammistes forsteri, *Bl. Schn.* p. 191.

Sparus pantherinus, *Lacép.* iv. p. 160, pl. 6. f. 1.

Cirrhites pantherinus, *Cuv. & Val.* iii. p. 70; *Less. Voy. Coqu. Poiss.* p. 225, pl. 22. f. 1.

Serranus tankervillæ, *Benn. Fishes of Ceylon,* pl. 27.

Cirrhites forsteri, *Günth. Fish.* ii. p. 71.

Great Comoro. *East coast of Africa and Cape to Pacific.*

Family SCORPÆNIDÆ.

SCORPÆNA, *Artedi.*

147.   **Scorpæna diabolus.**   [635.]

Scorpæna diabolus, *Cuv. & Val.* iv. p. 312; *Richardson, Ichth. Voy. Sulph.* p. 70, pl. 40; *Günth. Fish.* ii. p. 117.

Zanzibar. *East-Indian Ocean and Archipelago. Otaheiti.*

* Undetermined specimen [112].

148.   Scorpæna cirrhosa.   [34.]

Perca cirrhosa, *Thunberg, Nya Handl. Stockh.* xiv. 1793, p. 190, pl. 7. f. 2.
Scorpæna cirrhosa, *Cuv. & Val.* iv. p. 318; *Schleg. Fauna Japonica, Poiss.* p. 42, pl. 17. f. 2, 3; *Günth. Fish.* ii. p. 120.

D. 11 | $\frac{1}{9-10}$.   A. $\frac{3}{5}$.   L. lat. 50–58.

Aden. *East-Indian, Chinese, and Japanese seas.*

149.   Scorpæna longicornis, sp. n.   Plate VIII. fig. 1.   [570.]

D. 11 | $\frac{1}{9}$.   A. $\frac{3}{5}$.   L. lat. 36.

Palatine teeth. The height of the body is nearly equal to the length of the head, and is one-third of the total length. The head is nearly entirely scaleless. The jaws are equal in front. *The orbital tentacles are very long*, scarcely fringed, subcylindrical, *extending, when laid back, to the fourth dorsal spine.* The length of the snout is two-sevenths that of the head; interorbital space deeply concave, with scarcely any ridges within its concavity. Vertex with a deep subquadrangular groove, much broader than long, surrounded by spines.

The fifth and sixth dorsal spines are the longest, their length being contained twice and one-third in that of the head; the third is much stronger than the second of the anal, which is half the length of the head. There are five pectoral rays branched. Six longitudinal series of scales between the dorsal fin and the lateral line.

Colour reddish, marbled with white and black; a black blotch on the back, beneath and in front of the first four dorsal spines. Axil of pectoral marbled with light brown.

Length 2¼ inches.

Zanzibar.

150.   Scorpæna zanzibarensis, sp. n.   Plate VIII. fig. 2.   [400.]

D. 11 | $\frac{1}{9}$.   A. $\frac{3}{5}$.   L. lat. 37.

Palatine teeth. Height of body somewhat less than the length of the head, which is contained four times and one-third in the total length. Head nearly entirely scaleless; snout short, its length being about a quarter of that of the head. Interorbital space deeply concave, with scarcely any ridges at the bottom. Orbital tentacles well developed, reaching, when laid back, to the first dorsal spine; the part before the root of the pectorals scaly. Four pectoral rays branched. The fourth, fifth, and sixth dorsal spines the longest, their length being contained twice and two-thirds in the length of the head. The third dorsal spine much shorter than the second of anal, which is half the length of the head. Six longitudinal series of scales between the base of the dorsal fin and the lateral line, which has tentacles.

Colour brown, marbled with darker and lighter. The caudal with two light cross bands. The axil of pectoral marbled with light brown. The dorsal without any black spot.

Length 4½ inches.

Zanzibar.

PTEROIS, *Cuv.*

151.   **Pterois volitans.**   [516.]

Gasterosteus volitans, *Linn. Syst. Nat.* i. p. 491.
Pterois volitans, *Cuv. & Val.* iv. p. 352, pl. 88; *Günth. Fish.* ii. p. 122.

Zanzibar. *From East Africa, through Indian seas, to Australia.*

152.   **Pterois lunulata.**   [33.]

Pterois lunulata, *Schleg. Faun. Japon.* p. 45, pl. 19; *Günth. Fish.* ii. p. 124.

Aden. Zanzibar. *Japan.*

153.   **Pterois cincta.**   [501.]

Pterois cincta, *Rüpp. N. W. Fische*, p. 108, t. 26. f. 3; *Günth. Fish.* ii. p. 125.

Island of Johanna. *Red Sea. Navigator and Pearl Islands.*

154.   **Pterois brachyptera.**   [65, 206.]

Pterois brachyptera, *Cuv. & Val.* iv. p. 368; *Günth. Fish.* ii. p. 126.

$$D. 12 \mid \tfrac{1}{ij}. \quad A. \tfrac{3}{5-6}. \quad L. \text{ lat. } 45.$$

Zanzibar. *Seas of Ceram and Amboyna.*

TETRAROGE, *Günther.*

155.   **Tetraroge binotata.**   [320, 609.]

$$B. 7. \quad D. \tfrac{15}{8-9}. \quad A. \tfrac{3}{5-6}. \quad V. 1/5.$$

Apistus binotatus, *Peters in Wiegm. Arch.* 1855, p. 241.
Tetraroge binotata, *Günth. Fish.* ii. p. 134.

The height of the body is contained four times and one-sixth, and the length of the head five times in the total length. The upper profile of the snout descends rather vertically, the lower ascends in a gentle curve. Cleft of mouth oblique; the upper maxillary extends to a vertical from the anterior third of the eye. Præorbital spine very long and strong, reaching nearly to the base of the præopercular spine, which is somewhat shorter. Four teeth on the præopercular margin, two spines on operculum. Scales minute. Dorsal commences above front margin of eye, very high anteriorly; the last part of the spinous portion nearly even, and lower than the soft portion. The first dorsal spine is about one-third of the second, which is the longest and much longer than the head; the following three decrease in length; the sixth is the shortest; the remainder are very little longer. The dorsal is connected with the caudal by a slight membranous fold. The pectorals have twelve branched rays, and reach to the vent. The anal begins beneath the first dorsal ray, and is much lower than the dorsal; it has three spines, the third being longest, and two-thirds of the length of the head. The ventrals are connected with the body by a membranous fold.

Colour uniform brown, fins marked with darker and lighter. Dr. Peters mentions a spot on the lateral line, below the seventh and eighth dorsal spines, which is not visible in our specimen.

Length 5¼ inches.

No. [609] is an immature specimen 2⅓ inches long, in which the præorbital and præopercular spines are very minute, the second and third dorsal spines proportionately shorter, and scarcely any trace of scales is visible. The colour is lighter, with darker longitudinal lines.

Kiswara Bay, east coast of Africa. Zanzibar. *Mozambique.*

SYNANCEIA, *Schneider.*

156.   **Synanceia verrucosa.**   [340.]

Synanceia verrucosa, *Bl. Schn.* p. 195, t. 45; *Rüpp. N. W. Fische,* p. 109; *Günth. Fish.* ii. p. 146.
Scorpæna brachio, *Lacép.* iii. pp. 259, 272, pl. 12. f. 1.
Synanceia brachio, *Cuv. & Val.* iv. p. 447; *Cuv. Règne Anim. Ill. Poiss.* pl. 25. f. 3.

Zanzibar. *Red Sea. Indian Ocean and Archipelago. Polynesia.*

PLATYCEPHALUS, *Schneider.*

157.   **Platycephalus insidiator.**   [114.]

Cottus insidiator, *Forsk.* p. 25.
Platycephalus insidiator, *Bl. Schn.* p. 59; *Cuv. & Val.* iv. p. 227; *Rüpp. N. W. Fische,* p.102; *Schleg. Faun. Japon. Poiss.* p. 39, pl. 15. f. 1; *Günth. Fish.* ii. p. 177.

Aden. Zanzibar. *Red Sea. From east coast of Africa, through Indian Ocean, to Australia.*

158.   **Platycephalus pristis.**   [489.]

Platycephalus pristis, *Peters, Wiegm. Arch.* 1855, p. 240; *Günth. Fish.* ii. p. 188.

Zanzibar. *Mozambique.*

DACTYLOPTERUS, *Lacép.*

159.   **Dactylopterus orientalis.**   [124.]

Russell, pl. 161.
Dactylopterus orientalis, *Cuv. & Val.* iv. p. 134, pl. 76; *Schleg. Faun. Japon. Poiss.* p. 37, pl. 15 A; *Günth. Fish.* ii. p. 222.

Aden. Zanzibar. *Indian Ocean and Archipelago.*

Family TEUTHIDIDÆ.

TEUTHIS, *L.*

160. **Teuthis corallina.** [703.]

Amphacanthus corallinus, *Cuv. & Val.* x. p. 139; *Müll. & Schleg. Verh. Overz. Bezitt. Vissch.* p. 10, pl. 2. f. 2.
Teuthis corallina, *Günth. Fish.* iii. p. 316.

Seychelles. *Molucca Sea.*

161. **Teuthis stellata.** [275, 488.]

Scarus stellatus, *Forsk.* p. 26. no. 10.
Amphacanthus stellatus, *Bl. Schn.* p. 209; *Rüpp. N. W. Fische,* p. 129.
—— punctatus, *Rüpp. Atl. Fische,* p. 46, pl. 11. f. 2 (not *Bl. Schn.*).
—— nuchalis, *Cuv. & Val.* x. p. 140.
Teuthis stellata, *Günth. Fish.* iii. p. 320.

Zanzibar. *Red Sea. Ceylon.*

162. **Teuthis rostrata.** Plate X. fig. 2. [630.]

? Teuthis rostratus, *Cuv. & Val.* x. p. 158.

$$\text{D. } \tfrac{13}{10}. \quad \text{A. } \tfrac{7}{9}. \quad \text{V. } \tfrac{1}{4}.$$

The height of the body is contained thrice and a half in the total length, the length of the head five times and a half in the same; caudal deeply forked, the length of one of its central rays being two-fifths of that of the outer ones. The soft portion of the cheek, between the orbit, the præopercular limbs, and the snout, is lower than the orbit, and twice as long as high. There are from sixteen to seventeen teeth on each side of the upper and lower jaws. The anterior edge of the orbit is scarcely serrated. The spines of the fins are of moderate length and strength, the fifth of the dorsal being not quite half as long as the head, and a little longer than the snout, which is somewhat more produced than in the other species.

Ground-colour bluish, with numerous yellow vermiculated streaks broken up into spots on the anterior part of the body*; spinous portions of the fins brown, rays of the dorsal and anal with broad brown cross bands. Caudal lobes with several indistinct brownish cross bands.

Length 10 inches.
Zanzibar. ? *Red Sea.*

* As these disappear in preserved specimens, they are not shown in the figure.

163.   Teuthis nebulosa.   Plate X. fig. 3.   [26.]

Amphacanthus nebulosus, *Quoy & Gaim. Voy. Uran. Zool.* p. 369; *Cuv. & Val.* x. p. 164.
—— maculosus, *Quoy & Gaim. l. c.* p. 379.
—— gymnoparcius, *Richards. Ann. & Mag. Nat. Hist.* 1843, xi. p. 174.
—— olivaceus, *Cuv. & Val.* x. p. 163.
Teuthis nebulosa, *Günth. Fish.* iii. p. 321.

$$D. \frac{13}{10}. \quad A. \frac{7}{9}.$$

The height of the body is one-third of the total length, and the length of the head one-fourth of that without caudal. The snout is not much longer than the eye. The soft portion of the cheek, between the orbit, præopercular limbs, and snout, is two-thirds as high as long. There are twelve teeth on each side of the upper and lower jaws. The anterior margin of the orbit is obtusely denticulated. The spines of the fins are of moderate length and strength, the fifth of the dorsal being half as long as the head. Caudal fin deeply emarginate.

Body and vertical fins greyish olive, irregularly marbled with darker. Caudal with inconspicuous dark cross bands.

Length 6 inches.

Aden. Zanzibar. Island of Johanna. *Mauritius. Australia.*

Family BERYCIDÆ.

MYRIPRISTIS, *Cuv. & Val.*

164.   Myripristis murdjan.   [83, 207.]
Sciæna murdjan, *Forsk.* p. 48.
Myripristis murdjan, *Cuv. & Val.* iii. p. 177, vii. p. 495; *Günth. Fish.* i. p. 21.

Var. *a.* Red, each scale with a reddish silvery centre; opercular membrane deep red; fins red.
Var. *b.* Resembles *M. parvidens* (Cuv. & Val. iii. p. 173). It differs from var. *a* in having the denticulations of the upper maxillary very feeble, in the scapulary being entire, and in the more feeble striæ on the interoperculum.
Zanzibar. Aden. *Red Sea. Amboyna.*

165.   Myripristis adustus.   [191, 452.]
Myripristis adustus, *Bleek. Nat. Tydschr. Ned. Ind.* 1853, *Amboyna*, iii. p. 18; *Günth. Fish.* i. p. 22.

$$D. 10 \frac{1}{14\text{-}15}. \quad A. \frac{4}{11\text{-}13}. \quad L. lat. 28\text{-}30. \quad L. transv. 3/7.$$

Zanzibar. *Amboyna.*

H 2

HOLOCENTRUM, *Artedi*.

166.   **Holocentrum rubrum.**   [45, 67, 408.]
Sciœna rubra, *Forsk.* p. 48.
Holocentrum alborubrum, *Lacép.* iv. p. 372.
Holocentrus ruber, *Rüpp. Atlas, Fische*, p. 83, t. 22. f. 1.
Holocentrum rubrum, *Günth. Fish.* i. p. 35.
—— melanospilos, *Bleek. Acta Soc. Sci. Indo-Neerl.* iii. *Amboyna*, ix. p. 2.

Var. *a.* The specimens marked 45 and 67 correspond with those previously described as *H. rubrum*.

Var. *b*, marked [408], has a large black blotch at the base of the soft dorsal and anal, a third at the root of the caudal, and a fourth at the axil of the pectoral. This is clearly identical with that described by Bleeker as *H. melanospilos*.

These varieties are structurally identical, and the difference in coloration is probably a sexual one; one specimen of var. *b* proved to be a male fish, at or near spawning time.

Zanzibar.   Aden.   *Red Sea.   Indian archipelago.   Chinese and Japanese seas.*

167.   **Holocentrum spiniferum.**   [648.]
Sciœna spinifera, *Forsk.* p. 49.
Holocentrum leo, *Cuv. & Val.* iii. p. 204.
Holocentrus spinifer, *Rüpp. N. W. Fische*, pp. 96, 97, t. 25. f. 1 (not *Atlas*).
Holocentrum spiniferum, *Günth. Fish.* i. p. 39; *Kner, Novara, Fisch.* p. 7.

Chagos archipelago.   *Red Sea.   East coast of Africa, through the Indian Ocean, to the Pacific.*

168.   **Holocentrum diadema.**   [127, 192.]
Holocentrus diadema, *Lacép.* iv. pp. 372, 374, iii. pl. 32. f. 3; *Rüpp. Atlas, Fische*, p. 84, t. 22. f. 2.
Holocentrum diadema, *Cuv. & Val.* iii. p. 213; *Less. Voy. Coqu. Zool.* ii. p. 220, pl. 25. f. 2; *Günth. Fish.* i. p. 42.

Aden.   Zanzibar.   Island of Johanna.   *Red Sea.   Madagascar.   Ceylon.   Chinese seas.   Indian archipelago.*

169.   **Holocentrum sammara.**   [77, 453.]
Sciœna sammara, *Forsk.* p. 48.
Holocentrum sammara, *Rüpp. Atlas, Fische*, t. 22. f. 3; *Günth. Fish.* i. p. 46.

Island of Johanna.   *Red Sea.   Cape of Good Hope.   Amboyna.   Sumbawa.*

PEMPHERIS, *Cuv. & Val.* [*]

170.   **Pempheris mangula.**   [555.]
Mangula kutti, *Russell*, ii. pl. 114.
Pempheris mangula, *Cuv. & Val.* vii. p. 304; *Günth. Fish.* ii. p. 509.

Zanzibar.   *Indian Ocean and Archipelago.*

[*] Undetermined specimen [277].

Family POLYNEMIDÆ.

POLYNEMUS, *L.*

171. **Polynemus sexfilis.** [671.]

Polynemus sexfilis, *Cuv. & Val.* vii. p. 515; *Günth. Fish.* ii. p. 325.
Zanzibar. *Mauritius.*

172. **Polynemus sextarius.** [59.]

Polynemus sextarius, *Bloch,* p. 18, t. 4; *Cuv. & Val.* iii. p. 388, vii. p. 514; *Bleek. Verhand. Bat. Gen.* 1849, vol. xxii., *Perc.* p. 59; *Cant. Mal. Fish.* p. 32; *Günth. Fish.* ii. p. 326.
Bagamoia, east coast of Africa. *East Indian seas.*

Family SCIÆNIDÆ.

SCIÆNA, *Artedi.*

173. **Sciæna sina.** [143.]

Johnius sina, *Cuv. & Val.* v. p. 122.
Sciæna sina, *Günth. Fish.* ii. p. 292.
Bagamoia, east coast of Africa. Bombay. *Malabar. Japan.*

OTOLITHUS, *Cuv.*

174. **Otolithus argenteus.** [90.]

Otolithus argenteus, (*Kuhl & v. Hass.*) *Cuv. & Val.* v. p. 62; *Günth. Fish.* ii. p. 310; *Kner, Novara, Fisch.* p. 135, t. 6. f. 4 (air-bladder).
Aden. Mouth of Pangani river, east coast of Africa. *Seas of India, Ceylon, and Java.*

Family XIPHIIDÆ.

HISTIOPHORUS, *Lacép.*

175. **Histiophorus brevirostris, sp. n.** [682.]

The anterior portion of the dorsal is somewhat elevated, but much less so than in any species hitherto described, the third spine being considerably less than the height of the body; the next nine decrease rapidly in length, the ninth being equal to the first. All the succeeding spines are very short, none of them being equal in length to twice the diameter of the orbit. The rays of the second dorsal are longer than any of the spines after the ninth; the last is produced and reaches to the middle of the space between the dorsal and caudal. The second anal fin is similar to the second dorsal; the first is somewhat elevated, but less so than the first dorsal. The ventrals consist of one ray each, about half the length of the pectoral. The caudal lobes are contained five times in the total length.

The height of the body is contained once and five-sixths in the length of the head, and six times and a half in the total length. The upper jaw is somewhat compressed, broader than deep; its length to the nostrils is about three-fifths of that of the head. The produced part is rather longer than the distance from the point of the mandible to the posterior margin of the orbit. The upper profile of the head is nearly straight.

Colour uniform, without any spots or markings.

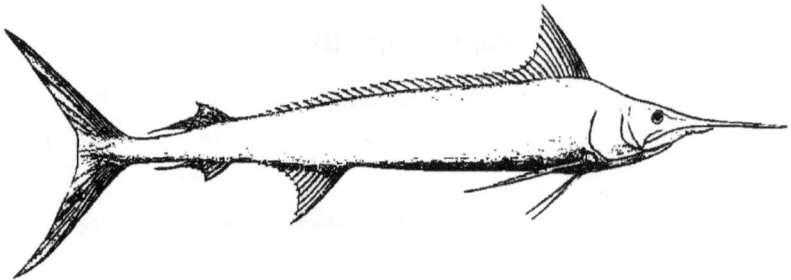

	ft.	in.
Total length	. 10	4
Greatest height	1	8
Length of head	. 3	0
· Length of produced part of jaw	1	1
Length of upper jaw from angle of mouth .	2	4
Width between base of pectorals . . .	0	10
Length of pectorals	1	11
Length of ventrals	0	9
Greatest height of dorsal . . .	1	2
Length of caudal lobes . . .	2	0

A single specimen of this fish was obtained by Colonel Playfair at Zanzibar as he was on the point of leaving that island for a few weeks; it was hurriedly stuffed and photographed before his departure; and it is from this photograph that the figure is taken. Unfortunately, during his absence the specimen was destroyed. The description is from notes taken immediately after the fish was captured.

There exists in the Museum of the Royal College of Surgeons in London a rostrum and portion of a skull of a *Histiophorus*, marked 251A, which most probably belongs to the same species. We subjoin a figure and a statement of the dimensions of this skull. It was obtained by purchase, but nothing further is known of its history.

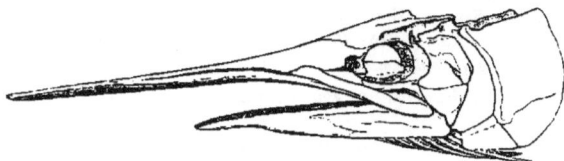

	ft.	in.
Distance from posterior margin of operculum to extremity of rostrum	4	2
Distance between extremity of maxillary and extremity of rostrum	3	2
Distance from anterior edge of orbit to extremity of rostrum	2	8
Length of maxillary	1	1
Length of produced part of upper jaw	1	5
Length of mandible	1	10
Depth of præoperculum	0	8¾

Family TRICHIURIDÆ.

TRICHIURUS, *L.*

176.   Trichiurus haumela. $\left[\frac{188}{A}\right]$

Clupea haumela, *Forsk.* p. 72.

Russell, pl. 41.

Trichiurus haumela, *Bl. Schn.* p. 518; *Cuv. & Val.* viii. p. 249; *Rüpp. N. W. Fische*, p. 41; *Cantor, Mal. Fish.* p. 113; *Günth. Fish.* ii. p. 348.

Bagamoia, east coast of Africa. *Indian Ocean and Archipelago.*

Family ACRONURIDÆ.

ACANTHURUS, *Schn.*

177. **Acanthurus triostegus.** [338.]

Seba, iii. 25. 4; *Russell*, i. pl. 86.
Chætodon triostegus, *L. Syst. Nat.* i. p. 463.
Acanthurus triostegus, *Bl. Schn.* p. 215; *Cuv. & Val.* x. p. 197; *Günth. Fish.* iii. p. 327.
—— hirudo, *Benn. Fishes of Ceylon*, pl. 11.

Mozambique. *From Mauritius to Polynesia and New Zealand.*

178. **Acanthurus matoides.** [18, 575.]

Acanthurus matoides, *Cuv. & Val.* x. p. 201; *Günth. Fish.* iii. p. 330.
—— annularis, *Cuv. & Val.* x. p. 209 (immature specimens with a whitish caudal ring).
—— blochii, *Cuv. & Val.* x. p. 209 (immature specimens without a whitish caudal ring).
—— xanthopterus, *Cant. Mal. Fish.* p. 202, pl. 4.

$$D. \frac{9}{26-28}. \quad A. \frac{3}{24-24}. \quad V. 1/5.$$

Var. *a* [18]. Body neutral-tint coloured, a broad bluish or yellow band across the base of caudal, a blue line along the base of dorsal, and an indistinct yellowish band from below the eye across forehead. Dorsal and anal brown, with three or four longitudinal blue bands, sometimes absent on the spinous portion of the former; anal sometimes with a blue margin. Exterior half of pectorals yellow or brownish.

Aden. Zanzibar. Length 9½ inches.

Var. *b* [575]. Colour brown, with indistinct yellow undulating lines; posterior margin of branchiostegous membrane black; a yellow band from orbit to orbit above nostril; several irregular yellow and blue lines on the snout; a yellowish band across base of caudal; a narrow yellow line along base of dorsal. Dorsal yellowish brown, with a blue basal line; beyond this there is a series of blue spots, round on the spinous, and elongate on the soft portion; above these there are several longitudinal bluish lines and a bright-blue margin. Ventrals longitudinally striped, alternately yellowish and bluish; pectorals bluish black inferiorly and yellowish superiorly.

Zanzibar. *East-Indian seas. Polynesia.* Length 9¼ inches.

179. **Acanthurus leucosternon.** [365.]

Acanthurus leucosternon, *Benn. Proc. Zool. Soc.* 1832, p. 183; *Bleek. Nat. Tydschr. Ned. Ind.* 1856, *Butoe*, iii. p. 237; *Günth. Fish.* iii. p. 340.
—— delisianus, *Cuv. & Val.* x. p. 193; *Guér. Iconogr. Poiss.* pl. 35. f. 2.

Zanzibar. *Mauritius. Ceylon. Butoe.*

180.   **Acanthurus ctenodon.**   [493, 585.]

Acanthurus ctenodon, *Cuv. & Val.* x. p. 211, pl. 289; *Günth. Fish.* iii. p. 342.
—— strigosus, *Bleek. Nat. Tydschr. Ned. Ind.* iv. p. 264, and vi. p. 102.

$$\text{D. } \frac{8}{21-30}. \quad \text{A. } \frac{3}{23-27}.$$

Var. *a* [585] agrees with the specimens in the British Museum, described by Dr. Günther (*l. c.*), especially with those from Amboyna.

Var. *b* [493] is of a dark-brown colour, covered with minute pearly spots, which disappear shortly after death.

Zanzibar. Ceylon. East-Indian archipelago. New Guinea. New Hebrides. Caroline Islands.

181.   **Acanthurus rhombeus.**   [278.]

Acanthurus flavescens, *Benn. Zool. Journ.* iv. 1828, p. 40.
—— rhombeus, *Kittl. in Mus. Senckenb.* i. p. 196, t. 13. f. 6; *Günth. Fish.* iii. p. 342.
—— scopas, *Cuv. & Val.* x. p. 245, pl. 290.
—— altivelis, *Cuv. & Val.* x. p. 249.

Colour uniform brown, with a greyish band descending obliquely from the suprascapula to the middle of the side.

Zanzibar. *From Mauritius to Polynesia. Coast of Cuba.*

182.   **Acanthurus xanthurus.** Plate VIII. fig. 4.   [116.]

Acanthurus xanthurus, *Blyth in Kelaart's Prodr. Faun. Ceylan. Append.* p. 50; *Günth. Fish.* iii. p. 343.

$$\text{D. } \frac{5}{21-25}. \quad \text{A. } \frac{3}{20-21}.$$

Height of the body rather more than two-thirds of the total length without caudal; snout rather produced; about eight lobate incisors on each side of the upper jaw; dorsal and anal fins rather elevated; scales minute, rough.

Colour wholly black, with a golden-yellow tail; pectorals yellowish.

Aden. *Coast of Ceylon.*

183.   **Acanthurus velifer.**   [9.]

Acanthurus velifer, *Bloch,* ix. p. 106, t. 427. f. 1; *Günth. Fish.* iii. p. 344.
—— blochii, *Benn. in Proc. Zool. Soc.* 1835, p. 207.

Zanzibar. *Sea of Batoe.*

184.   **Acanthurus desjardinii.**   [182.]

Acanthurus desjardinii, *Benn. in Proc. Zool. Soc.* 1835, p. 127; *Günth. Fish.* iii. p. 344.

Zanzibar. *Mauritius.*

This may prove to be only the adult state of *Acanthurus velifer*, Bloch.

I

NASEUS, *Commers.*

185.   **Naseus brevirostris.**   [195, 232.]

Naseus brevirostris, *Cuv. & Val.* x. p. 277, pl. 291; *Günth. Fish.* iii. p. 349.

$$D. \frac{5-6}{25-29}. \quad A. \frac{2}{25-30}.$$

Zanzibar. *Mauritius to Polynesia.*

186.   **Naseus tuberosus.**   [368.]

Naso tuberosus, *Lacép.* iii. p. 111, pl. 7. f. 3.
Acanthurus nasus, *Shaw, Zool.* v. p. 376, pl. 51.
Naseus tuber, *Cuv. & Val.* x. p. 290.
—— tuberosus, *Günth. Fish.* iii. p. 353.

Zanzibar. *From Mauritius to Polynesia.*

187.   **Naseus lituratus.**   [668.]

Harpurus lituratus, *Forst. Descr. Anim. ed. Licht.* p. 218.
Aspisurus elegans, *Rüpp. All. Fische,* p. 61, t. 16. f. 2.
Naseus lituratus, *Cuv. & Val.* x. p. 282; *Günth. Fish.* iii. p. 353.

$$D. \frac{6}{28-31}. \quad A. \frac{3}{21-31}.$$

Zanzibar. Island of Pemba. *Red Sea to Polynesia.*

Family CARANGIDÆ.

CARANX, *Cuv.*

188.   **Caranx rottleri.**   [92.]

Scomber rottleri, *Bl.* x. p. 40, t. 316.
Russell, ii. p. 33, pl. 143.
Caranx rottleri, *Rüpp. Atlas, Fische,* p. 102, and *N. W. Fische,* p. 48, 52; *Cuv. & Val.* ix. p. 29; *Günth. Fish.* ii. p. 421.

Aden. Zanzibar. *From the Red Sea through the Indian seas.*

189.   **Caranx kurra.**   [93.]

Kurra wodagawah, *Russell,* ii. p. 30, pl. 139.
Caranx russellii, *Rüpp. Atlas, Fische,* p. 99.
—— kurra, *Cuv. & Val.* ix. p. 44; *Günth. Fish.* ii. p. 427.
Decapterus kurra, *Bleek. Verh. Bat. Gen.* xxiv. *Makr.* p. 50, and *Nat. Tydschr. Ned. Ind.* 1851, p. 358.

Aden. Zanzibar. *Red Sea. Coast of Coromandel. Batavia.*

190.   **Caranx hasseltii.**   [699.]

? Caranx affinis, *Rüpp. N. W. Fische,* p. 49, t. 14. f. 1.
Saler hasseltii, *Bleek. Nat. Tydschr. Ned. Ind.* 1851, *Makreel.* p. 359 ; *Günth. Fish.* ii. p. 430.

D. 7–8 | $\frac{1}{25}$. A. 2 | $\frac{1}{20-21}$. L. lat. 45.

There are no detached rays in the fins; but the last of the dorsal and anal are somewhat remote from the penultimate rays, and slightly elongated and penicillated.
Seychelles. *Red Sea. East-Indian archipelago.*

191.   **Caranx djeddaba.**   [640.]

Scomber djeddaba, *Forsk.* p. 56.
Caranx djeddaba, *Rüpp. Atlas, Fische,* p. 97, pl. 25. f. 3 ; *Cuv. & Val.* ix. p. 51; *Günth. Fish.* ii.
  p. 432.

Zanzibar. *Red Sea. Pondicherry. Penang.*

192.   **Caranx vomerinus, sp. n.** Plate X. fig. 1.   [696.]

D. 7–8 | $\frac{1}{23}$. A. 2 | $\frac{1}{19}$. L. lat. 16.

Diagnosis.—Extremely similar to *C. leptolepis,* Cuv. & Val., but with a triangular patch of vomerine teeth; a narrow band of very minute teeth in the upper jaw, and generally in the lower; those of the tongue in an elongated band, but sometimes absent; a narrow band along each palatine bone.

The height of the body is two-sevenths of the total length; the length of the head somewhat less than one-fourth of the same. The lower jaw is very deep and projects much beyond the upper. The maxillary reaches nearly to the vertical from the front margin of the eye. Breast scaly. Lateral line slightly arched, becoming straight behind the middle of the soft dorsal; plates low, but forming a conspicuous keel. Base of caudal with a pair of distinct keels.

Colour.—Uniform white; a black opercular spot; the posterior margin of the præoperculum black.

Description.—The body is compressed and elliptical, with the upper and lower profiles equally curved; its greatest depth is below the fifth dorsal spine, where it is two-sevenths of the total length. The head is somewhat longer than high, and is contained four times and a quarter in the total length. The snout is longer than the diameter of the eye, which is two-sevenths of the length of the head; the cleft of the mouth is oblique, the lower jaw very prominent, the maxillary does not quite reach the vertical from the anterior margin of the orbit. The teeth are very minute, in narrow bands in the upper, and generally in the lower jaw; there is a triangular patch on the vomer, narrow bands on the palatines, and generally an elongated oval patch on the tongue. The

I 2

anterior and posterior margins of the eye are furnished with narrow adipose lids. The inferior margin of the præoperculum is rather shorter than the posterior one, both measurements being taken from the middle of the angle; they would form nearly a right angle were they not rounded at the junction. The parietal crest is very prominent, forming a sharp edge.

The spinous dorsal commences at about the vertical from the middle of the ventrals; its spines are very feeble, the second is longest, rather shorter than the. first ray, and about one-third of the length of the head; the two last are very minute and hardly project beyond the dorsal groove. The origin of the second dorsal is about midway between the snout and the end of the lateral line; with the exception of the first five, the rays of the soft dorsal are subequal in length. There is a very minute recumbent spine in front of the first dorsal. The caudal is deeply forked, the lobes being equal and about one-fifth of the total length. The soft portion of the anal is similar to the soft dorsal, and originates below the sixth ray of the latter; the detached spines are short, stout, and nearly equal. The ventral is inserted immediately behind the base of the pectorals, and extends beyond the vent. The pectoral is very long and falciform, and extends to the commencement of the straight portion of the lateral line; its length is about one-third of the total.

The scales are very small, covering the breast; the lateral line is slightly arched, becoming straight below the fifteenth dorsal ray; the straight portion is slightly armed with sixteen spiniferous plates, and it is continued in front of the spring of the arch as a row of minute pores which forms the chord of the arched portion. There are two very distinct keels on each side of the root of the caudal.

Length 8¾ inches.

Seychelles.

193. Caranx malabaricus. [697, 698.]

Scomber malabaricus, *Bl. Schn.* p. 31.
Russell, vol. ii. p. 37, pl. 150.
Caranx cæruleo-pinnatus, *Rüpp. Atl. Fische*, p. 100, and *N. W. Fische*, p. 47, t. 13. f. 2 (not *Cuv.*).
—— malabaricus, *Günth. Fish.* ii. p. 436.

Seychelles. *From the Red Sea, through all Indian seas, to the coast of Australia.*

194. Caranx bajad. [117, 647.]

Scomber bajad, *Forsk.* p. 55.
Caranx bajad, *Rüpp. Atlas, Fische*, p. 98, t. 25. f. 5; *Günth. Fish.* ii. p. 438.
—— fulvoguttatus, *Cuv. & Val.* ix. p. 75 (not *Rüpp.*).

Seychelles. *East coast of Africa. Red Sea. Batavia.*

No. 647 is probably a young specimen of *C. bajad*, with which it corresponds as to the radii; it is remarkable as having an outer series of stronger teeth in both jaws, and

some stronger ones on the vomer. The præoperculum is coarsely denticulated behind and below; the pectoral is not falciform. The colour is silvery below, darker above, with five or six indistinct darker cross bands and a black opercular spot.
Length 2 inches.

195. **Caranx speciosus.** [47.]

Scomber speciosus, *Forsk.* p. 54.
Russell, ii. p. 36, pl. 149.
Caranx speciosus, *Cuv. & Val.* ix. p. 130; *Cant. Mal. Fish.* p. 133; *Peters, Wiegm. Arch.* 1855, p. 245 (not *Rüpp.*); *Günth. Fish.* ii. p. 444.

Aden. Zanzibar. *From the Red Sea, through all the Indian seas, to the coast of New Holland.*

196. **Caranx rüppellii.** [700.]

Caranx petaurista, *Rüpp. Atl. Fische,* p. 95, pl. 25. f. 2 (not *Geoffr.*).
Caranx rüppellii, *Günth. Fish.* ii. p. 445.

Seychelles. *Red Sea.*

197. **Caranx melampygus.** [301.]

$$D. 8 \mid \tfrac{1}{23-25}. \quad A. 2 \mid \tfrac{1}{19}. \quad L. \text{ lat. } 36.$$

Caranx melampygus, *Cuv. & Val.* ix. p. 116; *Günth. Fish.* ii. p. 446.

Zanzibar. *Mauritius to Polynesia.*

198. **Caranx hippos.** [118.]

For synonymy see *Günther, Fish.* ii. p. 449.

Aden. Zanzibar. *Atlantic coasts of temperate and tropical America. All the Indian seas, to the Pacific and the shores of Australia.*

199. **Caranx armatus.** [74, 681, 428.]

Sciæna armata, *Forsk.* p. 53.
Russell, pl. 151. p. 38 (young).
Citula ciliaria, *Rüpp. Atl. Fische,* p. 102, t. 25. f. 8.
—— armata, *Rüpp. Atl. Fische,* p. 102, and *N. W. Fische,* p. 50.
Caranx citula, *Cuv. & Val.* ix. p. 126.
—— armatus, *Cuv. & Val.* ix. p. 127; *Cant. Mal. Fish.* p. 131; *Günth. Fish.* ii. p. 453.
—— ciliaris, *Cuv. & Val.* ix. p. 129 (young).
—— cirrhosus, *(Ehrenb.) Cuv. & Val.* pl. 250.
Olistus malabaricus, *Cuv. & Val.* ix. p. 137, pl. 251; *Cuv. Règne Anim. Ill. Poiss.* pl. 58. f. 1.
—— rüppellii, *Cuv. & Val.* ix. p. 144.

Aden. Zanzibar. Seychelles. *Red Sea, through all Indian seas, to New Guinea.*

200. <center>**Caranx ciliaris.**</center> [234.]

Zeus ciliaris, *Bloch*, vi. p. 29, t. 191.
Blepharis indicus, *Cuv. & Val.* ix. p. 154; *Schleg. Faun. Japon. Poiss.* p. 113, pl. 60. f. 3.
—— fasciatus, *Rüpp. Atl. Fische*, p. 129, t. 33. f. 2 (not *Richards.*).
Caranx ciliaris, *Günth. Fish.* ii. p. 454.

Cape Gardafui. Zanzibar. *From Red Sea through all Indian seas.*

201. <center>**Caranx gallus.**</center> [110.]

Zeus gallus, *L. Syst. Nat.* i. p. 454; *Bloch*, t. 192. f. 1.
Russell, i. p. 45, pl. 57, p. 46, pl. 58.
Gallichthys major, *Cuv. & Val.* ix. p. 168, pl. 254.
Scyris indica, *Rüpp. Atl. Fische*, p. 128, t. 33. f. 1, and *N. W. Fische*, p. 51; *Cuv. & Val.* ix. p. 145, pl. 252; *Cant. Mal. Fish.* p. 134.
Caranx gallus, *Günth. Fish.* ii. p. 456.

Aden. East coast of Africa. *Red Sea through all Indian seas.*

SERIOLA, *Cuv.*

202. <center>**Seriola nigro-fasciata.**</center> [369.]

Nomeus nigro-fasciatus, *Rüpp. Atl. Fische*, p. 82, t. 24. f. 2.
Seriola binotata, *Cuv. & Val.* ix. p. 215; *Cant. Mal. Fish.* p. 137.
—— nigro-fasciata, *Rüpp. N. W. Fische*, p. 51; *Günth. Fish.* ii. p. 465.

Zanzibar. *Red Sea. East-Indian Ocean and Archipelago.*

SERIOLICHTHYS, *Bleek.*

203. <center>**Seriolichthys bipinnulatus.**</center> [695.]

The Yellowtail (*Bennett*).

Seriola bipinnulata, *Quoy & Gaim. Voy. Uran. Zool.* i. p. 363, pl. 61. f. 3; *Cuv. Règne Anim. Ill. Poiss.* p. 130; *Jenyns, Zool. Beagle, Fishes*, p. 720.
Elagatis bipinnulatus, *Benn. Whaling Voyage*, ii. p. 283.
Seriolichthys bipinnulatus, *Bleek. Nat. Tydschr. Ned. Ind.* vi. p. 196; *Günth. Fish.* ii. p. 468.

D. 6 | $\frac{1}{26}$. I. A. $\frac{2}{16}$. I. Br. 7.

Quoy and Gaimard describe this fish as smaller than a mackerel, and measuring only 9 inches. There are two specimens in the British Museum, one from Madras, measuring 13 inches, and the other from Amboyna, 4½ inches in length. Colonel Playfair obtained two specimens at Seychelles, each of which measured 29 inches.

Bleeker's statement, that there are two præanal spines separate from the remainder of that fin, requires further confirmation. We are unable to detect any trace of them in the specimens we have examined.

There is one remarkable peculiarity regarding this fish which does not appear to have been observed. The second dorsal has a narrow scaly band along the base of the anterior portion. It is continuous with the line of the back as far as the seventh ray, when it takes an upward course, leaving a clear space between it and the line of the back; it diminishes gradually in length till it stops before the upper third of the fifteenth ray. The anal has a similar band, which terminates before the eighth ray.

Seychelles. *Amboyna. Madras.*

NAUCRATES, *Cuv.*

204. **Naucrates ductor.** [691.]

For synonymy see *Günther, Fish.* vol. ii. pp. 374, 469.

It has already been maintained by Gill and Kner that *Nauclerus,* Cuv. & Val., comprises the young specimens of *Naucrates ductor,* an opinion which, from the examination of several specimens in the British Museum and one from the east coast of Africa, proves to be quite correct. The armature of the præoperculum, which has been used by Cuvier for the distinction of various species, is found in very small examples only; it is very apparent in specimens an inch long, while it has disappeared in others of 2 inches. This armature is evidently subject to slight variations, and it is not impossible that all the six species of *Nauclerus* distinguished by Cuvier are referable to one or, at the outside, to two species of *Naucrates.*

A second question arises, whether *Naucrates* should be referred to *Scombridæ* or *Carangidæ.* The two anal spines of the young are separate from the soft portion; and although the number of caudal vertebræ is increased by two, yet the number of abdominal vertebræ remains the same: we are therefore inclined to remove this genus from the *Scombridæ* to the *Carangidæ.*

The specimen from the east coast of Africa corresponds entirely with Cuvier's description of *Nauclerus abbreviatus.* The caudal keels are not developed.

East coast of Africa. *All the seas of temperate and tropical regions.*

CHORINEMUS, *Cuv. & Val.* *

205. **Chorinemus lysan.** [639.]

Scomber lysan, *Forsk.* no. 67. pl. 54.

Russell, ii. p. 31, pl. 141.

Lichia lysan, *Rüpp. Atl. Fische,* p. 91.

Chorinemus commersonianus, *Cuv. & Val.* viii. p. 370.

—— lyzan, *Cuv. & Val.* viii. p. 387; *Rüpp. N. W. Fische,* p. 44; *Cantor, Mal. Fish.* p. 118.

—— lysan, *Günth. Fish.* ii. p. 471; *Kner, Novara, Fisch.* p. 163.

Zanzibar. *From the Red Sea to the north-west coast of Australia.*

* Undetermined specimen [260].

206.   **Chorinemus sancti Petri.**   [32, 167.]

Chorinemus sancti Petri, *Cuv. & Val.* viii. p. 379, pl. 236; *Peters, Wiegm. Arch.* 1855, p. 245; *Rüpp.
N. W. Fische*, p. 45; *Günth. Fish.* ii. p. 473.

Aden. West coast of India. Zanzibar. *All Indian seas.*

TRACHYNOTUS, *Lacép.*

207.   **Trachynotus ovatus.**   [375.]

Gasterosteus ovatus, *L. Syst. Nat.* i. p. 490.
Mookalee-parah, *Russell*, ii. p. 39, pl. 151.
For remainder of synonymy see *Günth. Fish.* ii. 481.

Zanzibar. *Tropical and temperate America. Both coasts of Africa. All Indian
seas to Australia.*

208.   **Trachynotus baillonii.**   [130, 131, 245.]

Russell, ii. pl. 142.
Cæsimorus baillonii, *Lacép.* iii. p. 93, pl. 3. f. 1.
—— quadripunctatus, *Rüpp. Atl. Fische*, p. 90, pl. 24. f. 1.
Trachynotus baillonii, *Günth. Fish.* ii. p. 484.

Aden. Zanzibar. *Indian Ocean.*

PSETTUS, *Commers.*

209.   **Psettus argenteus.**   [119, 468.]

Chætodon argenteus, *Linn. Amæn. Acad.* iv. p. 249.
Scomber rhombeus, *Forsk.* p. 58.
Psettus rhombeus, *Cuv. & Val.* vii. p. 215; *Peters in Wiegm. Arch.* 1855, p. 247.
—— argenteus, *Richards. Voy. Ereb. & Terr. Fishes,* p. 57, pl. 35. f. 1–3; *Günth. Fish.* ii. p. 487.

Aden. Zanzibar. *From the Red Sea, through all Indian seas, to the coasts of
Australia and Polynesia.*

PLATAX, *Cuv. & Val.*

210.   **Platax vespertilio.**   [226, 376.]

Chætodon vespertilio, *Bl.* tab. 199. f. 2; *Bennett, Ceylon Fishes*, pl. 5.
Platax vespertilio, *Cuv. Règne Anim.*; *Rüpp. Atl. Fische*, p. 143, and *N. W. Fische*, p. 33; *Cantor,
Mal. Fish.* p. 166; *Peters Wiegm. Arch.* 1855, p. 247; *Günth. Fish.* ii. p. 489.
—— albipunctatus, *Rüpp. Atl. Fische*, p. 69, t. 18. f. 4 (young).

The specimens marked [226] are adults, and that numbered [376] is an immature
specimen, and exactly corresponds with Rüppell's figure of *P. albipunctatus.*
Aden. Zanzibar. *From eastern coasts of Africa, through all Indian seas.*

211. Platax teira. [85.]

Chætodon teira, *Forsk.* p. 60, t. 22 ; *Bl.* t. 199. f. 1.

Platax teira, *Cuv. Règne Anim.* ; *Rüpp. Atl. Fische,* p. 68, and *N. W. Fische,* pp. 33, 37 ; *Cuv. & Val.* vii. p. 226 ; *Cantor, Mal. Fish.* p. 168 ; *Günth. Fish.* ii. p. 492.

Aden. Zanzibar. *East-Indian and Chinese seas.*

ZANCLUS, *Commers.*

212. Zanclus cornutus. [309.]

Chætodon cornutus, *L. Syst. Nat.* p. 461 ; *Bl.* t. 200. f. 2.

Zanclus cornutus, *Cuv. & Val.* vii. p. 102, pl. 177 ; *Günth. Fish.* ii. p. 493.

Zanzibar. *Indian Ocean and Archipelago. Polynesia.*

EQUULA, *Cuv.*

213. Equula edentula. [61.]

Scomber edentulus, *Bl.* t. 428 (bad).

Russell, i. p. 50, pl. 63.

Equula edentula, *Günth. Fish.* ii. p. 498.

Aden. Zanzibar. *Red Sea. Indian and Australian seas.*

GAZZA, *Rüpp.*

214. Gazza equulæformis. [261.]

Gazza equulæformis, *Rüpp. N. W. Fische,* p. 4, t. 1. f. 3 ; *Cantor, Mal. Fish.* p. 153 ; *Günth. Fish.* ii. p. 506.

Zanzibar. *Red Sea and East-Indian seas.*

Family STROMATEIDÆ.

STROMATEUS, *Artedi.*

215. Stromateus niger. [650.]

Russell, i. p. 32, pl. 43.

Stromateus niger, *Bloch,* xii. p. 93, t. 422 ; *Cuv. & Val.* ix. p. 385 ; *Cantor, Mal. Fish.* p. 139 ; *Günth. Fish.* ii. p. 401.

Apolectus stromateus, *Cuv. & Val.* viii. p. 439, pl. 238 ; *Cant. Mal. Fish.* p. 123.

Bagamoia, east coast of Africa. *East-Indian seas.*

K

Family CORYPHÆNIDÆ.

CORYPHÆNA, *Artedi.*

216.　　　　　　Coryphæna hippurus.　　　　　　[370.]

Coryphæna hippurus, *L. Syst. Nat.* p. 446; *Bl.* t. 174; *Cuv. & Val.* ix. p. 278, pl. 266; *Günth. Fish.*
ii. p. 405.
Lampugus pelagicus, *Cuv. & Val.* ix. p. 318 (immature).
Zanzibar.　*Mediterranean.　Atlantic.　Indian Ocean.　Pacific.*

MENE, *Lacép.*

217.　　　　　　Mene maculata.　　　　　　[66.]

Zeus maculatus, *Bl. Schn.* p. 95, pl. 22.
Russell, i. p. 47, pl. 60.
Mene maculata, *Cuv. & Val.* x. p. 104, pl. 285; *Schleg. Faun. Japon. Poiss.* p. 127, pl. 67. f. 3; *Günth.
Fish.* ii. p. 415.
Aden.　Zanzibar.　*East-Indian seas.*

Family NOMEIDÆ.

NOMEUS, *Cuv.*

218.　　　　　　Nomeus gronovii.　　　　　　[679.]

Gobius, sp., *Gronov. Zoophyl.* p. 82. no. 278.
Gobius gronovii, *L. Gm.* p. 1205.
Nomeus maculosus, *Benn. Proc. Comm. Zool. Soc.* i. 1831, p. 146.
—— mauritii, *Cuv. & Val.* ix. p. 243.
Seriola argyromelas, *Cuv. & Val.* ix. pl. 202.
Nomeus gronovii, *Günth. Fish.* ii. p. 387.

Mozambique Channel.　*Tropical parts of Atlantic.　Indian Ocean and coasts of
Australia.*

Family SCOMBRIDÆ.

SCOMBER, *Artedi.*

219.　　　　　　Scomber microlepidotus.　　　　　　[70, 252.]

Scomber microlepidotus, *Rüpp. N. W. Fische,* p. 38, t. 11. f. 2; *Cantor, Mal. Fish.* p. 105; *Günth.
Fish.* ii. p. 361.
Zanzibar.　Aden.　*Red Sea.　Penang and China.*

THYNNUS, *Cuv. & Val.*

220. **Thynnus thunnina.** [430.]

Thynnus thunnina, *Cuv. & Val.* viii. p. 104, pl. 212; *Schleg. Faun. Japon. Poiss.* p. 95, pl. 48; *Günth. Fish.* ii. p. 364.

Seychelles. *Mediterranean. Tropical parts of Atlantic. East-Indian seas.*

221. **Thynnus pelamys.** [722.]

Scomber pelamys, *L. Syst. Nat.* i. p. 492.
Thynnus pelamys, *Cuv. & Val.* viii. p. 113, pl. 214; *Schleg. Faun. Japon. Poiss.* p. 96, pl. 49; *Günth. Fish.* ii. p. 365.

Open sea, east coast of Africa. *Tropical parts of Atlantic and Indian Ocean.*

AUXIS, *Cuv. & Val.*

222. **Auxis rochei.** [660.]

Scomber rochei, *Risso, Ichth. Nice,* p. 165.
Auxis vulgaris, *Cuv. & Val.* viii. p. 139, pl. 216.
—— rochei, *Günth. Fish.* ii. p. 369.

Zanzibar. *Mediterranean. Tropical parts of Atlantic. East-Indian archipelago.*

CYBIUM, *Cuv.* *

223. **Cybium commersonii.** [415.]

Scomber commersonii, *Lacép.* ii. p. 600, pl. 20. f. 1.
Russell, pl. 135.
Cybium commersonii, *Cuv. Règne Anim.; Rüpp. Atlas, Fische,* p. 94, t. 25. f. 1; *Cuv. & Val.* viii. p. 163; *Günth. Fish.* ii. p. 370.

Zanzibar. *From east coast of Africa, through all Indian seas.*

ELACATE, *Cuv.*

224. **Elacate nigra.** [94.]

Gasterosteus canadus, *L. Syst. Nat.* p. 491.
Scomber niger, *Bl.* t. 337.
Russell, ii. p. 39, pl. 153.
Elacate pondiceriana, *Cuv. & Val.* viii. p. 329; *Rüpp. N. W. Fische,* p. 43, t. 12. f. 3.
—— motta, *Cuv. & Val.* viii. p. 332.
—— malabarica, *Cuv. & Val.* viii. p. 332.

* Undetermined specimen [146].

K 2

Elacate atlantica, *Cuv. & Val.* viii. p. 33+, pl. 233.
—— bivittata, *Cuv. & Val.* viii. p. 338; *Schleg. Faun. Japon. Poiss.* p. 104, pl. 56; *Cant. Mal. Fish.* p. 116.
—— nigra, *Günth. Fish.* ii. p. 375.

Zanzibar. Aden. *Indian seas. Tropical parts of the Atlantic.*

ECHENEIS, *Artedi.*

225.   **Echeneis remora.**

For synonymy see *Günth. Fish.* ii. p. 378.

Aden. Zanzibar. *Seas of temperate and tropical regions.*

226.   **Echeneis naucrates.**   [106.]

Echeneis naucrates, *L. Syst. Nat.* i. p. 446; *Bl.* ii. p. 131, t. 171; *Cantor, Mal. Fish.* p. 199; *Schleg. Faun. Japon. Poiss.* p. 270, pl. 120. f. 1; *Richards. Ichth. China,* p. 203; *Günth. Fish.* ii. p. 384.

Aden. Zanzibar. *Seas of temperate and tropical regions.*

Family TRACHINIDÆ.

PERCIS, *Schneid.*

227.   **Percis punctulata.**   [721.]

Percis punctulata, *Cuv. & Val.* iii. p. 265; *Günth. Fish.* ii. p. 238.

Zanzibar. *Mauritius.*

228.   **Percis polyophthalma.**   [506.]

Percis polyophthalma, (*Ehrenb.*) *Cuv. & Val.* iii. p. 272.

Zanzibar. *Red Sea. Indian Ocean.*

229.   **Percis hexophthalma.**   [225.]

Percis hexophthalma, (*Ehrenb.*) *Cuv. & Val.* iii. p. 271, vii. p. 507; *Günth. Fish.* ii. p. 239.
—— cylindrica, *Rüpp. Atlas, Fische,* p. 19, t. 5. f. 2 (not *Cuv.*).

This and the preceding species have been united into one by some authors; but we entertain doubts as to the propriety of doing so. All the specimens with three ocellated spots on each side (*P. hexophthalma*) have about ten narrow oblique brown lines radiating from the lower part of the eye, over the opercles, whilst the specimens with six or seven ocelli on each side (*P. polyophthalma*) have the sides of the head dotted. Unfortunately the internal organs of our specimens are not sufficiently well preserved to enable us to determine the sexes, so as to ascertain whether the differences between them are sexual.

One specimen of the *hexophthalma* form is a male, two appear to have been sterile; and one of the *polyophthalma* form was probably a female.

Zanzibar. *Red Sea. Indian Ocean. Louisiade archipelago.*

230.   **Percis xanthozona.**   Plate VIII. fig. 3.   [447.]

Percis xanthozona, *Bleek. Verhand. Bat. Genootsch.* 1849, vol. xxii. p. 59; *Günth. Fish.* ii. p. 210.

D. 5 | 21.   A. $\frac{1}{17-18}$.   L. lat. 63–65.   L. transv. 6/18.

The two middle dorsal spines are the longest, the second not being very much longer than the first. The height of the body is contained seven times and two-thirds, and the length of the head thrice and three-fourths in the total length. The interorbital space is very narrow, the bony bridge between the eyes being about one-third of their horizontal diameter. The ventrals do not quite, the pectorals do reach to the vent. The præoperculum is entire, the caudal convex.

The general colour of the upper portion of the body is light greyish olive, the lower part is white, these two colours being separated by a light lateral band. The upper lip has three dark-brown spots on each side; the upper surface of the head and body is marked with numerous brown dots, arranged in about eight broad but indistinct cross bands. Along the lower side of the body is a series of about ten large dark-brown spots, connected together at the upper surface by a narrower and lighter band; there is a single small spot in each interspace between the larger ones. The dorsal is very faint yellow, with three longitudinal series of black dots; the anal is white, with a single series, and a black and white margin. The caudal has a large milk-white patch and irregular black spots and markings.—Length 6 inches.

Zanzibar. *Sea of Batavia.*

231.   **Percis pulchella.**   [242.]

Percis pulchella, *Schleg. Faun. Japon. Poiss.* p. 24, pl. 10. f. 2; *Günth. Fish.* ii. p. 240.

Zanzibar. *Japan.*

SILLAGO, *Cuv.*

232.   **Sillago sihama.**   [180.]

Atherina sihama, *Forsk.* p. 70.

Russell, t. 113.

Sillago sihama, *Rüpp. Atlas, Fische,* p. 9, t. 3. f. 1; *Günth. Fish.* ii. p. 243.

Aden. Zanzibar. Seychelles. *All Indian seas.*

OPISTHOGNATHUS, *Cuv.*

233.   **Opisthognathus nigromarginatus.**   [333.]

Opisthognathus nigromarginatus, *Rüpp. Atlas, Fische,* t. 28. f. 4; *Günth. Fish.* ii. p. 254.

Zanzibar. Mozambique. *Red Sea. India.*

Family PEDICULATI.

ANTENNARIUS*, *Commers.*

234. **Antennarius marmoratus.** [418.]

For synonymy see *Günth. Fish.* iii. p. 185, also *Bleek. Atl. Ichth.* iv. p. 23, pl. 198. f. 4, and pl. 199.
f. 1.

Zanzibar. *Seas between the tropics.*

235. **Antennarius nummifer.** [102.]

Chironectes nummifer, *Cuv. Mém. Mus.* iii. p. 430, pl. 17. f. 4; *Cuv. & Val.* xii. p. 425; *Rüpp. N.W. Fische,* p. 141.
Antennarius nummifer, *Bleek. Nat. Tydschr. Ned. Ind.* 1854, *Amboina,* v. p. 497, and *Atl. Ichth.* iv.
p. 18, pl. 198. fig. 2; *Günth. Fish.* iii. p. 195.

D. 3 | 11–12. A. 7. P. 12.

Aden. *Red Sea. East-Indian seas.*

Family GOBIIDÆ.

It is remarkable that so many species of this family, which consists of fish inhabiting the shores of the sea and the mouths of rivers, and which are never met with in the open sea, should be found on both sides of the Indian Ocean. In some instances the Indian and African specimens are identical, and in others the variation between them is very slight indeed.

GOBIUS, *Artedi†.*

236. **Gobius giuris.** [303, 504.]

Russell, pl. 50, 51, 53.
Gobius giuris, *Buch. Ham. Fishes Ganges,* p. 51, pl. 33. f. 15; *Günth. Fish.* iii. p. 21.

From a comparison of several fully adult examples from East Africa, from 9 to 14 inches long, with specimens from India, we are enabled to confirm an opinion advanced in Dr. Günther's Catalogue of Fishes, iii. p. 21, that this species inhabits both sides of the Indian Ocean, and that *G. platycephalus,* Peters, is really identical with *G. giuris.* This species has been found by Col. Playfair far up rivers on the east coast of Africa and in the sea at Zanzibar.

Zanzibar. Pangani River. *All Indian coasts, entering rivers.*

* Undetermined specimen [419]. † Undetermined specimens [150, 304, 394].

237. **Gobius nebulo-punctatus.** [63, 386, 388.]

Gobius nebulo-punctatus, *Cuv. & Val.* xii. p. 58 ; *Peters, Wiegm. Archiv*, 1855, p. 250 ; *Günth. Fish.* iii. p. 26.
—— fuscus, *Rüpp. Atl. Fische,* p. 137.

Aden. Zanzibar. *Mozambique. Red Sea.*

238. **Gobius ophthalmotænia.** [406, 598.]

Gobius ophthalmotænia, *Bleek. Nat. Tydschr. Ned. Ind.* 1854, vii. *Kokos-Eiland.* p. 46 ; *Günth. Fish.* iii. p. 37.
—— capistratus, *Peters, Wiegm. Archiv,* 1855, p. 251.

Zanzibar. *Mozambique. East-Indian archipelago. Chinese seas.*

239. **Gobius caninus.** Plate LX. fig. 1. [371.]

Gobius caninus, *Cuv. & Val.* xii. p. 86 ; *Günth. Fish.* iii. p. 38.
—— grandinosus, *Val. in Voy. Bonite, Poiss.* p. 177, pl. 5. f. 4.

Varietas *africana*, Pl. LX. fig. 1. This Indian species occurs also on the east coast of Africa, though with slight variations, which, however, do not appear to be important enough to warrant the establishment of a distinct species. The African form appears to attain a larger size, specimens being found of from 5 to 6 inches in length. The dark violet spot on the shoulder of Indian specimens is here absent, and there are ten longitudinal series of scales between the second dorsal and anal. The upper two-thirds of the caudal also, in old specimens, is covered with blackish spots.

Zanzibar. *East-Indian archipelago. China.* Entering rivers.

240. **Gobius albomaculatus.** [48, 401.]

Gobius albomaculatus, *Rüpp. Atl. Fische,* p. 135, and *N. W. Fische,* p. 137 ; *Peters, Wiegm. Arch.* 1855, p. 250 ; *Günth. Fish.* iii. p. 69.
—— quinquocellatus, *Cuv. & Val.* xii. p. 95.

Aden. Zanzibar. *Red Sea.*

241. **Gobius sewardii,** sp. n. [446.]

$$D. 6 \mid \tfrac{1}{16}. \quad A. \tfrac{1}{16}.$$

Closely allied to *G. fontanesii,* Bleek. Caudal rather elongate, pointed, its length being contained four times and a third in the total length. The scales on the tail are rather larger than those on the trunk, which are minute. The height of the body is equal to the length of the head, and about one-sixth of the total length. The eyes are situated very close together ; their diameter is about a quarter of the length of the head. There is an outer series of larger teeth in both jaws, and a large canine on each side of the lower one. The cleft of the mouth is oblique, and the lower jaw is longer than the upper. The first dorsal is somewhat higher than the second; the spines are slightly produced and filiform; both fins are much lower than the body.

Colour pale, with light-blue spots above and a few transverse blue streaks below the lateral line; blue markings on cheek, opercles, and base of pectorals; vertical fins clouded with reddish.—Length 4½ inches.

Zanzibar.

Named after Dr. G. E. Seward, Surgeon to the Zanzibar Political Agency.

GOBIOSOMA, *Girard.*

242. **Gobiosoma fasciatum,** sp. n. [487.]

$$D. 6 \mid \tfrac{1}{10}. \quad A. \tfrac{1}{8}.$$

The height of the body is about one-fifth of the length without caudal, and the length of the head about a quarter of the same. The snout is very obtuse; there is a canine tooth on each side of the lower jaw. The first dorsal spine is rather produced.

Body brownish red, with four darker cross bands; the first across the back before the dorsal; the second below the spinous dorsal; the third below the third, fourth, and fifth dorsal rays; and the fourth below the end of the soft dorsal. The head has numerous small blue ocelli, edged with purplish. Dorsal and caudal fins yellowish; the soft portion of the former has about fourteen light-bluish darker-edged longitudinal lines; anal blackish brown, with seven blue longitudinal bands, each edged with purple; pectoral with numerous yellow dots; ventrals brownish black.—Length 5 inches.

Zanzibar.

GOBIODON *Kuhl & v. Hass.*

243. Gobiodon citrinus, [64.]

Gobiodon citrinus, *Rüpp. N. W. Fische,* p. 139, t. 32. f. 4; *Günth. Fish.* iii. p. 87.

Aden. *Red Sea.*

244. **Gobiodon reticulatus,** sp. n. Plate IX. fig. 2. [64 A.]

$$D. 6 \mid 12\text{--}13. \quad A. 9\text{--}11.$$

In habit similar to the other species of this genus; scales none; height of body two-fifths of the total length without caudal; length of the head one-fourth of the same; a pair of strong teeth at the symphysis of the lower jaw; eye small; dorsal and anal fins elevated, but lower than the body; the spines of the anterior dorsal decrease slightly in length posteriorly, and the last is connected with the first ray by a low membrane; ventral disk very short, not adherent to the belly.

Colour brownish olive; the posterior part of the trunk and tail covered with small round whitish spots, separated by a network of darker lines; dorsal and anal fins blackish, with a white black-edged band along the base; head uniform.

Length 1 inch 3 lines.

Aden.

PERIOPHTHALMUS, *Schn.*

245. **Periophthalmus koelreuteri.** [426.]

Periophthalmus koelreuteri, *Cuv. & Val.* xii. p. 181; *Rüpp. N. W. Fische,* p. 140; *Günth. Fish.* iii. p. 97.

Seychelles. Zanzibar. *Red Sea to Australia and the islands of the Western Pacific.*

ELEOTRIS, *Gronov.*

246. **Eleotris ophiocephalus.** [425.]

Eleotris ophiocephalus, (*Kuhl & v. Hass.*) *Cuv. & Val.* xii. p. 239; *Cant. Mal. Fish.* p. 196; *Günth. Fish.* iii. p. 107.

—— margaritacea, *Cuv. & Val.* xii. p. 240.

The African specimens do not differ from those of the East Indies, except in having the second dorsal and caudal fins spotted with dark brown instead of with white.

Fresh water of Seychelles and Johanna; salt water of Mozambique. *East-Indian archipelago. New Ireland. Philippines.*

247. **Eleotris madagascariensis.** [405 A.]

Eleotris madagascariensis, *Cuv. & Val.* xii. p. 210; *Günth. Fish.* iii. p. 111.

Mozambique. *Madagascar.*

248. **Eleotris butis.** [363.]

Cheilodipterus butis, *Buch. Ham.* pp. 57, 367; *Gray & Hardw. Ill. Ind. Zool.* ii. pl. 93. f. 3.

Eleotris humeralis, *Cuv. & Val.* xii. p. 246.

—— butis, *Cant. Mal. Fish.* p. 196; *Günth. Fish.* iii. p. 116.

The only difference between the African and Indian specimens is, that the scales of the former have the appearance of being slightly carinated.

Fresh water of Johanna. Salt water of Mozambique. *East Indies. East-Indian archipelago.*

249. **Eleotris wardii,** sp. n. Plate IX. fig. 3. [689.]

D. 6 | $\frac{1}{12}$. A. $\frac{1}{12}$. L. lat. ca. 95.

This species is allied to *E. periophthalmus* and to *E. muralis.* Head naked; the

L

height of the body is contained seven times and a half in the total length, and the length of the head four times and a half in the same. Head elongate, nearly twice as long as high, with the snout somewhat prominent and longer than the diameter of the eye. Lower jaw longest. Eyes close together, and a quarter of the length of the head. No spine at the angle of the præoperculum. Outer series of teeth slightly enlarged, with a small canine at each side of the lower jaw. Each dorsal spine is produced into a short filament; the last dorsal and anal rays are also somewhat produced. The caudal is acutely rounded.

Colour of the body pearly, with four broad brown cross bands, the anterior and posterior edges of which are darker: the first is below the five anterior dorsal spines; the second corresponds to the interspace between the dorsal fins; the third is below the end of the soft dorsal; and the fourth is across the caudal. There is a rudimentary band in the interspaces between each pair, the first being across the neck. A straight silvery blue band from the lower part of the maxillary to the upper angle of the operculum, and a faint blue line (disappearing after death) from the lower anterior margin of the eye to the middle of the intermaxillary. The first dorsal is brownish, with some white marks and a large black white-edged ocellus on the summit. The second dorsal is yellowish white, with a blackish band along the middle and two round black spots on its base. The anal is shaded with yellow, white, and pale brown, with a dark-brown margin.

Length 3½ inches.

Zanzibar.

We have named this species after Swinburne Ward, Esq., Her Majesty's Civil Commissioner for the Seychelles.

250. **Eleotris fusca.** [559, 642.]

Pœcilia fusca, *Bl. Schn.* p. 453.
Cheilodipterus cunnus, *Buch. Ham. Fish. Ganges*, p. 55, pl. 5 f. 16.
Eleotris nigra, *Quoy & Gaim. Voy. Uran.* p. 259, pl. 60. f. 2; *Cuv. & Val.* xii. p. 233.
—— mauritianus, *Benn. Proc. Comm. Zool. Soc.* i. p. 166.
Eleotris fusca, *Günth. Fish.* iii. p. 125.

Fresh water of Seychelles and Johanna. Pangani River, east coast of Africa. *Madagascar, Mauritius, and thence to Polynesia.*

251. **Eleotris soaresi,** sp. n. Plate IX. fig. 4. [405.]

D. 6 | $\frac{1}{8}$. A. $\frac{1}{8}$. L. lat. 62–63.

Very closely allied to *E. fusca* and to *E. melanosoma*; differing from the former in having the cheeks and lower half of the gill-apparatus entirely naked; it has a broader snout and a more prominent head; from the latter it is distinguished by smaller scales.

There are seventeen longitudinal series of scales between the origins of the dorsal and anal fins; the præoperculum is furnished with a spine directed downwards; the scales on the upper surface of the head are small, extending to between the eyes. The height of the body is contained five times and a half in the total length, the length of the head three times and a half in the same. The head is broad and depressed, as is also the snout; the lower jaw does not project much beyond the upper one. The diameter of the eye is about one-seventh of the length of the head, and three-fifths of the width of the interorbital space; the maxillary extends to below the middle of the eye. The teeth are arranged in villiform bands; the caudal is short and rounded.

Colour dark brown; dorsal and anal punctulated with black.—Length 5½ inches. Mozambique.

We have named this species after Sr. João da Costa Soares of Mozambique.

252.   **Eleotris microlepis.** Plate IX. fig. 5.   [612.]

Eleotris microlepis, *Bleek. Nat. Tydschr. Ned. Ind.* 1856, *Banda*, v. p. 102; *Günth. Fish.* iii. p. 132.
Eleotriodes microlepis, *Bleek. l. c.* 1858, *Coram*, p. 212.

$$D. 6 \mid \tfrac{1}{27}. \quad A. \tfrac{1}{31\text{-}27}.$$

Body elongate, slightly compressed; mouth very protractile, its cleft oblique, the lower jaw projecting beyond the upper. The height of the body is one-eighth, and that of the head one-sixth of the total length.

Several series of teeth in the upper jaw, with an outer series of larger ones; a single series on the sides of the lower jaw, several in front. Dorsal spines slender, flexible, the sixth somewhat remote from the others; the rays of the soft dorsal are simple and much higher than the body. The anal rays are branched and equal in height to the body. Caudal subtruncated; its length is contained five times and a half in that of the body. Scales minute, imbedded in the skin.

Colour in spirits.—Uniform light yellowish, with some wavy pearly bands on the sides of the head; *a deep-black streak across the base of the lower two-thirds of the pectoral rays.*

Colour in life.—Pale greenish, with a number of very faint flesh-coloured cross bands, those on the front part straight, the remainder crescent-shaped, the horns pointing forward; these cease before the tail, on which are about three similar longitudinal bands, which are continued on the caudal fin. Snout brilliant green. One or two blue wavy lines on the snout and opercles; base of pectorals blue, with a black crescent-shaped line. First dorsal with a green base, above which is a blue band; the rest of the fin is flesh-coloured. Second dorsal ray very faint orange; the rays, and a line between each pair, parallel to them, a little darker than the rest of the fin. Anal

almost colourless, with a blue and flesh-coloured band along the base.   Caudal greenish, with an upper flesh-coloured margin and a pale line between each pair of rays.
  Length 4¾ inches.
  Zanzibar.   *East-Indian archipelago.*

CALLIONYMUS, L.

253.   **Callionymus longecaudatus.**   [620 м, 674 ғ.]

Callionymus japonicus, *Houtt. Verh. Holl. Maatsch. Wet. Haarl.* xx. p. 311 (not *Val.*).
—— longicaudatus, *Schleg. Faun. Japon. Poiss.* p. 151, pl. 78. figs. 1, 2, pl. 79A. fig. 1, *Günth. Fish.* iii. p. 148.
—— reevesii, *Richards. Voy. Sulph. Fishes,* p. 60, pl. 36. figs. 1–3 (not fig. 4).
—— variegatus, *Schleg. Faun. Japon. Poiss.* p. 153 (fem.).
—— belcheri, *Richards. Voy. Sulph. Fishes,* p. 62, pl. 37. figs. 1, 2.
  Zanzibar.   *East-Indian archipelago.   China.   Japan.*

254.   **Callionymus marmoratus.**   [295 м & ғ.]

Callionymus marmoratus, *Peters, Wiegm. Arch.* 1855, p. 255; *Günth. Fish.* iii. p. 150.
—— perelegans, *Bianc. Spec. Zool. Mossamb.* p. 263, *Pisc.* fig. 9.
  Zanzibar.   *Mozambique.*

Family BLENNIIDÆ.

PETROSCIRTES, *Rüpp.*

255.   **Petroscirtes elongatus.**   [79, 472.]

Petroscirtes elongatus, *Peters, Wiegm. Arch.* 1855, p. 249; *Günth. Fish.* iii. p. 233.
  Zanzibar.   *Mozambique.*

256.   **Petroscirtes variabilis.**   [325.]

Petroscirtes variabilis, *Cant. Mal. Fish.* p. 200; *Günth. Fish.* iii. p. 234.
—— cynodon, *Peters, Wiegm. Arch.* 1855, p. 246.
  Zanzibar.   *Mozambique.   Penang.   Port Jackson.*

257.   **Petroscirtes barbatus.**   [465.]

Petroscirtes barbatus, *Peters, Wiegm. Arch.* 1855, p. 248; *Günth. Fish.* iii. p. 238.
  Zanzibar.   Mombassa.   *Mozambique.*

SALARIAS, *Cuv.*

258. **Salarias fasciatus.** [644.]

Blennius gattorugine, *Forsk.* p. 23.
—— fasciatus, *Bl.* ii. p. 111, t. 162. f. 1.
Salarias quadripinnis, *Rüpp. Atl. Fische,* p. 112, t. 28. f. 2; *Cuv. & Val.* xi. p. 318.
—— fasciatus, *Cuv. & Val.* xi. p. 324; *Günth. Fish.* iii. p. 244.
Island of Johanna. *Red Sea. Indian Ocean. Polynesia.*

259. **Salarias dussumieri.** Plate IX. figs. 6, 7. [62, 385, 391.]

Salarias dussumieri, *Cuv. & Val.* xi. p. 310; *Günth. Fish.* iii. p. 251.

D. 12 | 20–21. A. 22.

[391.] Adult males, three and a half inches long, want the orbital tentacles. The ground-colour of the lower half of the body is pearly; the body is crossed by seven or eight darker bands, the four anterior of which are broken up into a pair of bands each. The whole body, bands as well as interspaces, is covered with small dark-brown dots. The first dorsal is uniform blackish, the second dorsal and caudal are dotted with brown like the body, and the anal has one or two series of brown, white-edged dots along the middle.

[385.] The adult female, four and a half inches long, has a slightly fringed orbital tentacle as long as the eye. The body is marked with irregular pairs of dark-brown streaks and with a few scattered brown spots; the dots on the dorsal fins are confluent into oblique lines; the caudal and anal are immaculate.

[62.] Young females, one and a half to two and a half inches long, have cross bands on the body, but the fins are immaculate.

Aden. Zanzibar. *Cape of Good Hope. Coast of Malabar. Port Essington.*

260. **Salarias oorti.** [719 A.]

Salarias oorti, *Bleek. Natur. Tydschr. Ned. Ind.* i. p. 257, f. 15, and *Act. Soc. Sci. Indo-Neerl.* iii. *Sumatra,* p. 39; *Günth. Fish.* iii. p. 257.

As only one specimen, two and a half inches long, exists in this collection, it is somewhat doubtful whether the determination is correct.

Zanzibar. *East-Indian archipelago.*

261. **Salarias unicolor.** [719.]

Salarias unicolor, *Rüpp. N. W. Fische,* p. 136; *Günth. Fish.* iii. p. 259.

D. 12 | 17–18. A. 18–20.

In the specimens from the Mozambique coast the colour is greenish grey, with small white spots rather irregularly scattered over the sides and vertical fins (caudal excepted).

These fins are marked with indistinct oblique darker bands or lines, which are broadest on the first dorsal.

Mozambique coast. *Red Sea.*

Family SPHYRÆNIDÆ.

SPHYRÆNA, *Cuv.*

262. **Sphryræna jello.** [429.]

Jellow, *Russell*, pl. 174.

Sphyræna jello, *Cuv. & Val.* iii. p. 349; *Rüpp. N. W. Fische*, p. 98; *Cant. Mal. Fish.* p. 24; *Bleek. Verhand. Bat. Gen.* 1857, vol. xxvi. *Sphyræna*, p. 12; *Günth. Fish.* ii. p. 337.

Seychelles. *Red Sea. Cape of Good Hope. Bay of Bengal. East-Indian archipelago.*

263. **Sphyræna obtusata.** [100.]

Sphyræna obtusata, *Cuv. & Val.* vi. p. 350; *Cant. Mal. Fish.* p. 24; *Bleek. Verhand. Bat. Gen.* 1857, vol. xxvi. *Sphyræna*, p. 17; *Günth. Fish.* ii. p. 339.

Aden. Zanzibar. *Mauritius. East Indies. Port Jackson.*

264. **Sphyræna agam.** [707.]

Esox sphyræna, *Forsk.* p. 16.

Sphyræna agam, *Rüpp. N. W. Fische*, p. 99, t. 25. f. 2; *Günth. Fish.* ii. p. 341.

Zanzibar. *Red Sea.*

Family ATHERINIDÆ.

ATHERINA, *Artedi.*

265. **Atherina forskälii.** [393, 483.]

? Atherina hepsetus, *Forsk.* p. 69.

——— forskälii, *Rüpp. N. W. Fische*, p. 132, t. 33. f. 1; *Cant. Mal. Fish.* p. 103; *Günth. Fish.* iii. p. 397.

Zanzibar. *Red Sea. Penang.*

Family MUGILIDÆ.

MUGIL, *Artedi.*

266. **Mugil cæruleo-maculatus.** [86, 341.]

? Mugil cæruleo-maculatus, *Lacép.* v. pp. 385, 389; *Cuv. & Val.* xi. p. 128.

Mugil cæruleo-maculatus, *Bleek. Nat. Tydschr. Ned. Ind.* 1851, Riouw, p. 481, and *Act. Soc. Sci. Indo-Neerl.* viii., Sumatra, ix. p. 5; *Günth. Fish.* iii. p. 415.

Aden. Zanzibar. *Mauritius. East-Indian archipelago.*

267. **Mugil ceylonensis.** [293, 654, 714.]

Mugil ceylonensis, *Günth. Fish.* iii. p. 446.

Zanzibar.   Island of Johanna.   *Ceylon.*

AGONOSTOMA, *Benn.*

268. **Agonostoma telfairii.** [577.]

Agonostomus telfairii, *Benn. Proc. Comm. Zool. Soc.* 1830, p. 166.

Nestis cyprinoides, *Cuv. & Val.* xi. p. 167, pl. 317.

Agonostoma telfairii, *Günth. Fish.* iii. p. 462, and *Zool. Record,* ii. p. 192.

Fresh water of Johanna.   *Mauritius.   Réunion.*

Family FISTULARIDÆ.

FISTULARIA, *L.*

269. **Fistularia serrata.** [98.]

Russell, ii. p. 58, pl. 173.

Fistularia serrata, *Cuv. Règne Anim.* (after *Bloch*) ; *Günth. Fish.* iii. p. 533; *Kner, Novara, Fisch.* p. 238.

—— commersonii, *Rüpp. N. W. Fische,* p. 142 ; *Peters, Wiegm. Arch.* 1855, p. 258.

Cannorhynchus immaculatus, *Cant. Mal. Fish.* p. 211.

Aden.   Zanzibar.   Seychelles.   *Through the Indian Ocean to the seas of China and New Holland.*

AULOSTOMA, *Lacép.*

270. **Aulostoma chinense.** [603 A & B.]

Fistularia chinensis, part., *L. Syst. Nat.* i. p. 515.

Aulostoma chinense, *Schleg. Faun. Japon. Poiss.* p. 320 ; *Peters, Wiegm. Arch.* 1856, p. 258 ; *Günth. Fish.* iii. p. 538.

Var. *a* corresponds to var. *α* of *Günther.*

Var. *b* corresponds to var. *β* of *Günther* : in addition to the markings therein specified our specimen has several light spots between the dorsal and anal and on the tail.

Length 20 inches.

Zanzibar.   *From the coast of Mozambique to the Western Pacific.*

Family CENTRISCIDÆ.

CENTRISCUS, *L.*

271. **Centriscus gracilis.**

Centriscus gracilis, *Lowe, Proc. Zool. Soc.* 1839, p. 86, and *Trans. Zool. Soc.* iii. p. 12; *Günth. Fish.* iii. p. 521.

Two immature specimens were obtained in a tow-net on the east coast of Africa, near Zanzibar, measuring six lines in length.

East *and West* coasts of Africa. *Sea of Madeira. Coasts of Japan and China.*

AMPHISILE, *Klein.*

272. **Amphisile punctulata.** [36.]

Amphisile, sp., *Klein, Pisc. Miss.* iv. p. 28, t. 6. f. 6.

—— punctulata, *Bianconi, Spec. Zool. Mossamb.* fasc. x. 185t, p. 221, t. 1. f. 2; *Günth. Fish.* iii. p. 527.

—— brevispinis, *Peters, Wiegm. Arch.* 1855, p. 259.

—— punctata, *Kner in Sitzsber. Wien. Akad.* 1860, xxxix. p. 534, f. 2.

Aden. Zanzibar. *Red Sea. Mozambique.*

Family LABYRINTHICI.

OSPHROMENUS, *Commers.*

273. **Osphromenus olfax.** [420.]

For synonymy see *Günther, Fish.* iii. p. 382.

Naturalized in Seychelles and *Mauritius Fresh waters of East-Indian archipelago.*

Order ACANTHOPTERYGII PHARYNGOGNATHI.

Family POMACENTRIDÆ.

AMPHIPRION, *Schn.*

274. **Amphiprion bicinctus.** [201, 343.]

Amphiprion bicinctus, *Rüpp. Atl. Fische,* p. 139, t. 35. f. 1; *Cuv. & Val.* ix. p. 505; *Günth. Fish.* iv. p. 8.

a. D. $\frac{10}{16}$. A. $\frac{2}{14}$. Colour black, with two pale-blue cross bands; the anterior part of the head, thorax and belly, and anterior part of tail brilliant orange; posterior part of tail and caudal whitish, the other fins bright orange.

b. D. $\frac{11}{15}$ A. $\frac{2}{14}$. The general colour of this variety is darker and less brilliant than the previous one.

Length of each from 4½ to 5 inches.

Zanzibar. *Red Sea.*

DASCYLLUS, *Cuv.*

275. **Dascyllus aruanus.** [101.]

Chætodon aruanus, *L. Syst. Nat.* i. p. 464; *Bloch,* iii. p. 62, t. 198. f. 2.
—— abu dafur, *Forsk.* p. 15.
Pomacentrus aruanus, *Rüpp. Atl. Fische,* p. 39.
Dascyllus aruanus, *Cuv. & Val.* v. p. 434; *Günth. Fish.* iv. p. 12.
Chætodon araneus, *Benn. Fish. of Ceylon,* pl. 17.

Berbera, north-east coast of Africa. Zanzibar. *Thence to Polynesia and New Zealand.*

276. **Dascyllus trimaculatus.** [13, 485.]

Pomacentrus trimaculatus, *Rüpp. Atl. Fische,* p. 39, t. 8. f. 3.
Dascyllus trimaculatus, *Cuv. & Val.* v. p. 441; *Günth. Fish.* iv. p. 13.

Aden. Zanzibar. *From the east coast of Africa to Polynesia.*

277. **Dascyllus marginatus.** [10, 344.]

Pomacentrus marginatus, *Rüpp. Atl. Fische,* p. 38, t. 8. f. 2.
Dascyllus marginatus, *Cuv. & Val.* v. p. 439, pl. 133. f. 2; *Günth. Fish.* iv. p. 14.

Aden. Zanzibar. Ibo. *Red Sea.*

POMACENTRUS, *Lacép.**

278. **Pomacentrus annulatus.** [331.]

Pomacentrus annulatus, *Peters, Wiegm. Arch.* 1855, p. 265; *Günth. Fish.* iv. p. 18.

Coast of Mozambique.

279. **Pomacentrus pavo.** [264, 594.]

Chætodon pavo, *Bl.* t. 198. f. 1.
Pomacentrus pavo, *Rüpp. Atl. Fische,* p. 37; *Cuv. & Val.* v. p. 413; *Cuv. Règne Anim. Ill. Poiss.* pl. 32. f. 2; *Peters, Wiegm. Arch.* 1855, p. 265; *Günth. Fish.* iv. p. 23.

Zanzibar. Mozambique. *East-Indian archipelago.*

* Undetermined specimens [484, 646].

M

280. **Pomacentrus obtusirostris.** Plate X. fig. 4. [638.]

Pomacentrus obtusirostris, *Günth. Fish.* iv. p. 24.

D. $\frac{13}{11-13}$. A. $\frac{2}{11-14}$. L. lat. 30. L. trans. $\frac{3-1}{9}$.

The height of the body is two-sevenths of the total length; snout obtuse, much shorter than the diameter of the orbit, the width of which is more than that of the interorbital space. There are only six small incisors anteriorly in the upper jaw, and eight in the lower, the lateral teeth being very small. Præorbital very narrow and not serrated; operculum with two spines. The scales on the upper side of the head do not advance to the front margin of the orbit. Of the dorsal spines, the fourth, fifth, and sixth are the longest, the posterior ones being a little shorter; the soft portion of the anal and the caudal lobes are produced into points.

Colour pale bluish above, white below, the scales being marked with darker-blue elongated spots arranged so as to form longitudinal lines. A deep-black spot superiorly at the axil of the pectoral.

Length 2 to 4 inches.

Zanzibar.

281. **Pomacentrus trilineatus.** [49.]

Pomacentrus trilineatus, (*Ehrenb.*) *Cuv. & Val.* v. p. 428; *Günth. Fish.* iv. p. 25.
—— biocellatus, *Rüpp. N. W. Fische,* p. 127, t. 31. f. 3 (young, not good).

Aden. *Mozambique. Red Sea. Molucca Sea.*

282. **Pomacentrus bankanensis.** [15, 158, 508, 643.]

Pomacentrus tæniops, *Bleek. Nat. Tydschr. Ned. Ind.* 1852, Banka, ii. p. 729 (not *C. & V.*).
—— bankanensis, *Bleek. l. c.* 1853, Sumatra, iii. p. 513; *Günth. Fish.* iv. p. 26.

Aden. Zanzibar. Island of Johanna. *East-Indian archipelago. Sea of China.*

283. **Pomacentrus punctatus.** [402.]

Pomacentrus punctatus, *Quoy & Gaim. Voy. Uran. Zool.* p. 395, pl. 64. f. 1; *Rüpp. Atl. Fische,* p. 37;
Cuv. & Val. v. p. 429; *Günth. Fish.* iv. p. 29; *Kner, Novara, Fisch.* p. 242.

Zanzibar. *Mauritius. Red Sea. Bola Bola.*

GLYPHIDODON, *Lacép.*

284. **Glyphidodon cœlestinus.** [14, 212.]

Chætodon saxatilis, *Forsk.* p. 62; *Bloch,* t. 206. f. 2.
Russell, i. p. 67, pl. 86.
Chætodon tyrwhitti, *Benn. Fishes of Ceylon,* pl. 25.
Glyphisodon saxatilis, *Rüpp. Atl. Fische,* p. 35, and *N. W. Fische,* p. 126.

Glyphisodon rahti, *Cuv. & Val.* v. p. 456, ix. p. 507; *Cant. Mal. Fish.* p. 142.
—— cœlestinus, (*Soland.*) *Cuv. & Val.* v. p. 464, ix. p. 508.
Apogon quinquevittatus, *Blyth, Journ. As. Soc. Beng.* 1859, p. 272, and 1861, p. 111.
Glyphidodon cœlestinus, *Günth. Fish.* iv. p. 38.

· Aden. Zanzibar. Seychelles. *Red Sea to Polynesia.*

285. **Glyphidodon sordidus.** [539.]

Chætodon sordidus, *Forsk.* p. 62. no. 87.
Russell, pl. 85.
Glyphisodon sordidus, *Rüpp. Atl. Fische,* p. 34, t. 8. f. 1 ; *Cuv. & Val.* v. p. 466.
—— gigas, *Liénard, Dix. Rapp. Soc. Hist. Nat. Maur.* p. 35.
Glyphidodon sordidus, *Günth. Fish.* iv. p. 41.

Aden. Zanzibar. *Mauritius. Indian seas.*

286. **Glyphidodon sparoides.** [50, 455.]

Glyphisodon sparoides, *Cuv. & Val.* v. p. 468; *Peters, Wiegm. Archiv,* 1855, p. 266.
Glyphidodon sparoides, *Günth. Fish.* iv. p. 44.

Zanzibar. Island of Johanna. *Mauritius. Mozambique.*

287. **Glyphidodon melanopus.** [540.]

Glyphidodon melanopus, *Bleek. Nat. Tydschr. Ned. Ind.* xi. 1856, p. 82; *Günth. Fish.* iv. p. 48.

Zanzibar. *East-Indian archipelago.*

288. **Glyphidodon melas.** [184.]

Glyphisodon melas, (*Kuhl & v. Hass.*) *Cuv. & Val.* v. p. 472 ; *Günth. Fish.* iv. p. 45.
—— ater, *Cuv. & Val.* v. p. 473.

D. $\frac{13}{13-14}$. A. $\frac{2}{12-13}$. L. lat. 28. L. transv. $\frac{3\frac{1}{2}}{10}$.

Aden. Zanzibar. *Red Sea. East-Indian archipelago. New Hebrides.*

289. **Glyphidodon leucogaster.** [482.]

Glyphidodon leucogaster, *Bleek. Verh. Batav. Genootsch.* 1847, *Labr. Cten.* p. 26; *Günth. Fish.* iv. p. 46.

Zanzibar. *East-Indian archipelago.*

290. **Glyphidodon adenensis,** sp. n. Plate XI. fig. 1. [95.]

D. $\frac{13}{15}$. A. $\frac{2}{10}$. L. lat. 29 or 30. L. transv. $4\frac{1}{3}/14$.

The height of the body is contained once and two-thirds in the length without caudal ; the length of the head three times and a fifth in the same. The upper profile descends

м 2

rapidly from the dorsal fin towards the snout. Teeth rather narrow, about thirty-six in the upper jaw. Præorbital ring very narrow below the middle of the eye, its width being about one-third of the diameter of the orbit; breadth of the interorbital space rather more than the diameter of the eye. The fifth, sixth, and seventh dorsal spines longest, and longer than the last; soft dorsal higher than long; free portion of the tail very short and elevated.

Colour in a dried state uniform brownish olive, with darker lines along the series of scales. Vertical fins blackish; ventrals black; a black spot superiorly at the angle of the pectorals.

Length 5 inches.

Aden.

HELIASTES, *Cuv.*

291. **Heliastes lepidurus.** [712.]

Heliases lepidurus, *Cuv. & Val.* v. p. 498.
Heliastes lepidurus, *Günth. Fish.* iv. p. 63.

Zanzibar. East-Indian archipelago. New Guinea.

292. **Heliastes opercularis,** sp. n. Plate XI. fig. 2. [527.]

D. $\frac{13}{16}$. A. $\frac{2}{10}$. L. lat. 29. L. transv. 3/9.

Evidently closely allied to *H. axillaris*, Benn., and to *H. xanthurus*, Bleek. The height of the body is contained three times and a half in the total length, or twice and a half in that without caudal; the length of the head is one-fourth of the latter. Teeth in a narrow band, with an outer series of larger ones. Width of the præorbital less than half the diameter of the eye; breadth of the interorbital space greater than the diameter of the orbit; præoperculum distinctly emarginate on its posterior limb. Spines of dorsal slender, the fourth and fifth the longest, thence they decrease gradually in length, the last being equal to the second; soft portion much higher than long, one of the longest rays is once and two-thirds the length of the fourth dorsal spine. Caudal deeply forked. Second anal spine half the length of the head. Scales on the cheeks in four series; præorbital scaly.

Colour brownish, the cutaneous sheath of each scale with a darker margin. *A large irregular black patch behind the gill-opening, a similarly coloured line along the posterior margin of the præoperculum, a large black spot around base of pectorals, a smaller oblong one on superior surface of ventrals, and a still smaller irregular spot on the central caudal rays.* Dorsal and anal black, with light posterior margins; caudal brownish, with upper and lower edges black and posterior margins light.

Length 5⅔ inches.

Zanzibar.

Family LABRIDÆ.

XIPHOCHILUS, *Bleek*.

293. **Xiphochilus robustus.** Plate XII. fig. 3. [667.]

Xiphochilus robustus, *Günth. Fish.* iv. p. 98.

This species is sufficiently described in Günther's Catalogue of Fishes. In the specimen from Zanzibar the formula of the dorsal fin is $\frac{12}{8}$, and in that from Mauritius $\frac{11}{9}$; in the latter specimen the twelfth spine is turning into a ray, the lower half being spinous and the upper half articulated, but not branched.

Colour in life.—Upper part fawn, lower white. A broad whitish oblique band ascending from the root of the pectoral to the back of the tail behind the termination of the dorsal; on the upper margin of this band the colour of the body is darker; each scale, except on the band and lower part of the body, has an indistinct pale-blue vertical line. Lips cobalt-blue; a similarly coloured line through the centre of the orbit, another along its base, and a third below it; a blue line on the operculum, parallel to the posterior limb of the præoperculum ; opercular membrane blue. Five or six longitudinal blue lines on the tail. Dorsal yellowish, with blue margin and base and a band of blue marks along the middle. Anal yellowish, with a series of blue spots along base, a bluish margin, and a very distinct yellow intramarginal band. Caudal with blue lines along the rays ; ventrals and pectorals pale fawn.

Length 11 inches.

Zanzibar. *Mauritius.*

294. **Xiphochilus gymnogenys,** sp. n. Plate XII. fig. 4. [198.]

D. $\frac{12}{8}$. A. $\frac{3}{10}$. L. lat. 29. L. transv. $2\frac{1}{2}/11$.

Diagnosis.—The greater part of cheek and of sub- and interoperculum naked ; but there is a patch of small scales, as large as the eye, below its posterior edge. Hind part of gill-cover produced into a long flexible flap, fitting above the axil of the pectoral. Interoperculum extremely broad, flexible, partly covering its fellow below the throat.

Description.—*Xiphochilus robustus*, and still more *X. gymnogenys*, form a transition between this genus and *Chœrops*, the pharyngeal bones of *X. gymnogenys* bearing a much greater resemblance to that genus than to *X. typus*.

The height of the body is contained nearly four times, and the length of the head nearly three times and two-thirds in the total length. The maxillary does not reach to the anterior margin of the orbit. Præorbital much higher than the orbit. Cheek naked, with the exception of a patch of scales below the posterior margin of the eye. The posterior margin of the præoperculum very slightly denticulated ; the operculum, suboperculum, and the upper part of the interoperculum scaly; the last is as broad as the diameter of the eye. Dorsal spines stout, the longest being one-fifth of the length of the head; they increase in length posteriorly. Third anal spine longest; it is equal to the third of the dorsal, and is contained five times and a half in the length of the head. Caudal truncated. First ventral ray produced.

Colour rosy; a series of bright-blue spots on the back, close to the base of the dorsal, and two or three more or less indistinct longitudinal series of pearly spots along the sides. Axil of pectoral blue. A greenish blotch on the posterior part of the operculum, and a broad blue patch on the inferior half of the præoperculum. An indistinct blue band round the orbit, continued on the snout to about the middle of the maxillary. Lower lip bluish, with a very faint blue band parallel to it lower down.

Length 7 inches.

Zanzibar.

PTERAGOGUS, *Peters.*

295. **Pteragogus tæniops.** [266, 476.]

Cossyphus tæniops, *Peters, Wiegm. Arch.* 1855, p. 262.
Pteragogus tæniops, *Günth. Fish.* iv. p. 102.

$$\text{D. } \frac{10}{10}. \quad \text{A. } \frac{3}{9}. \quad \text{L. lat. 23–25.}$$

Very similar in general appearance to *Duymæria*, but differing from it in the formula of the fins; the first ray of the ventral also is much more produced than the lobes of the dorsal.

The height of the body is contained twice and two-thirds, and the length of the head four times in the total length. Two posterior canine teeth in the upper jaw. The vertical limb of the præoperculum distinctly serrated. The lobes of the anterior dorsal and anal spines are very slightly produced; the spines of the former increase in length posteriorly. The first ventral ray is much produced, and reaches to the middle of the anal. Caudal acutely rounded.

Colour in life violet, variegated with darker and lighter, and covered with numerous faint blue spots. A **Y**-shaped darker mark between anterior margins of orbits and front of snout; a similarly coloured band between the posterior margins of orbits, and a vertical band with lighter edges from base of orbits to posterior joint of lower jaw; a narrow blue line parallel to the denticulated limb of præoperculum; a rectangular, black, light-edged spot between the first and second dorsal spines; the rest of that fin is marbled with red, white, and brown, and has an orange margin. A black spot on base of pectorals.

Length 5½ inches.

Zanzibar. *Mozambique.*

296. Pteragogus opercularis. [576.]

Cossyphus opercularis, *Peters, Wiegm. Arch.* 1855, p. 261 (not *Guichenot*).
Pteragogus opercularis, *Günth. Fish.* iv. p. 101.

Zanzibar. Mozambique.

COSSYPHUS, *Cuv. & Val.*

297. Cossyphus axillaris. [280.]

Labrus axillaris, *Benn. Proc. Comm. Zool. Soc.* i. 1831, p. 166.
Cossyphus axillaris, *Cuv. & Val.* xiii. p. 131, pl. 371 ; *Günth. Fish.* iv. p. 103.

Zanzibar. *Mauritius. Madagascar. Ulea. New Hebrides.*

298. Cossyphus diana. [229, 409.]

Labrus diana, *Lacép.* iii. pp. 451, 522, pl. 32. f. 1.
Cossyphus diana, *Cuv. & Val.* xiii. p. 127 ; *Bleek. Atlas Ichth.* i. p. 150, t. 38. f. 1 ; *Günth. Fish.* iv. p. 104.
—— spilotes, *Guichenot, in Maillard, Notes sur l'île de la Réunion, Annexe C. Poiss.* p. 14.

Zanzibar. *Mozambique. Mauritius. East-Indian archipelago.*

299. Cossyphus bilunulatus. [228.]

Labrus bilunulatus, *Lacép.* iii. pp. 454, 526, pl. 31.
Cossyphus bilunulatus, *Cuv. & Val.* xiii. p. 121 ; *Bleek. Atl. Ichth.* i. p. 161, t. 38. f. 3 ; *Günth. Fish.* iv. p. 105.

Zanzibar. *Mauritius. East-Indian archipelago.*

300. Cossyphus atrolumbus. [490.]

Cossyphus atrolumbus, *Cuv. & Val.* xiii. p. 123 ; *Günth. Fish.* iv. p. 105.

Zanzibar. *Mauritius. West Pacific.*

LABROIDES, *Bleek.*

301. Labroides dimidiatus. [284.]

Labrus latovittatus, *Rüpp. N. W. Fische,* p. 7, t. 2. f. 2 (not *Lacép.*).
Cossyphus dimidiatus, *Cuv. & Val.* xiii. p. 136.
Labroides dimidiatus, *Bleek. Atl. Ichth.* i. p. 156, t. 44. f. 1 ; *Günth. Fish.* iv. p. 119.

Zanzibar. *Red Sea. East-Indian archipelago.*

302. Labroides quadrilineatus. [72.]

Labrus quadrilineatus, *Rüpp. N. W. Fische,* p. 6, t. 2. f. 1.
Cossyphus tæniatus, *Cuv. & Val.* xiii. p. 134.
—— quadrilineatus, *Cuv. & Val.* xiii. p. 135 ; *Günth. Fish.* iv. p. 120.

Aden. *Red Sea.*

DUYMÆRIA, *Bleek.*

303.   **Duymæria filamentosa.**   [279, 495, 496.]

Cossyphus filamentosus, *Peters, Wiegm. Arch.* 1855, p. 263.
Duymæria filamentosa, *Günth. Fish.* iv. p. 122.

$$\text{D. } \tfrac{9}{11}. \quad \text{A. } \tfrac{3}{9-10}. \quad \text{L. lat. 26.}$$

Var. *a* [279, 495]. Lobes of the anterior dorsal and anal spines much produced; the former sometimes higher than the body. Caudal rounded; the middle rays longest. Colour greenish, marbled with darker and lighter; a few darker spots along the lateral line; a black spot between the first and second dorsal spine.—Length 5 inches.
Aden. Zanzibar. *Mozambique.*

Var. *b* [496]. Colour greyish, variegated and spotted with black, and covered with brilliant cobalt spots disappearing in a dried state. Numerous fine blue lines radiating from the eye; a cluster of deep-black dots on each side of the nape of the neck, three of them being on the posterior margin of the orbit. A flesh-coloured band across middle of vertical limb of præoperculum and angle of operculum; another fainter one from angle of mouth. Produced lobes of dorsal black; the remainder of the fin yellowish. Caudal yellow, with two longitudinal rows of minute points, alternately black and pale blue, between each pair of rays. Anal yellowish; produced lobes pinkish.
—Length 5¼ inches.
Zanzibar.

CHEILINUS *, *Lacép.*

304.   **Cheilinus trilobatus.**   [308, 558.]

Cheilinus trilobatus, *Lacép.* iii. pp. 529, 537, pl. 31. f. 3; *Rüpp. Atl. Fische,* p. 22; *Cuv. & Val.* xiv. p. 79; *Bleek. Atl. Ichth.* i. p. 66, t. 27. f. 2; *Günth. Fish.* iv. p. 126.

Aden. Zanzibar. *From east coast of Africa to China and New Hebrides.*

305.   **Cheilinus mossambicus.**   [478, 510.]

? Cheilinus radiatus, *Cuv. & Val.* xiv. p. 91.
Cheilinus radiatus, *Peters, Wiegm. Arch.* 1855, p. 264 (not *Bl. Schn.*).
—— mossambicus, *Günth. Fish.* iv. p. 127.

$$\text{D. } \tfrac{9}{9-10}. \quad \text{A. } \tfrac{3}{8}. \quad \text{L. lat. 21–23.}$$

Zanzibar. *Mozambique. East-Indian archipelago. New Hebrides.*

306.   **Cheilinus undulatus.**   [621.]

Cheilinus undulatus, *Rüpp. N. W. Fische,* p. 20, t. 6. f. 2; *Cuv. & Val.* xiv. p. 108; *Bleek. Atl. Ichth.* p. 68, t. 26. f. 3; *Günth. Fish.* iv. p. 129.

* Undetermined specimen [35].

Although there cannot be the slightest doubt that the specimen from Zanzibar is identical with that described and figured by Bleeker, we are by no means certain that it is identical with the specimen from the Red Sea described by Rüppell. First, Rüppell makes no mention of the black longitudinal ocular stripes, which are so very distinct, even in dried specimens; secondly, Rüppell represents his specimen as having the cheek-scales in three series, while there are only two in the Zanzibar and Bleeker's specimens; and thirdly, the fish figured by Rüppell appears to be much deeper in the body. Nevertheless all these specimens may belong to one and the same species; but we thought it right to direct attention to the points mentioned above.

Zanzibar. Batavia. ? Red Sea.

307. <p style="text-align:center">**Cheilinus fasciatus.**</p> [347.]

Sparus fasciatus, *Bl.* v. p. 18, t. 257.
Cheilinus fasciatus, pt., *Rüpp. Atl. Fische,* p. 23.
—— fasciatus, *Cuv. & Val.* xiv. p. 92; *Rüpp. N. W. Fische,* p. 18; *Bleek. Atl. Ichth.* p. 67, t. 26. f. 2; *Günth. Fish.* iv. p. 129.

Zanzibar. From Red Sea, through all Indian seas.

308. <p style="text-align:center">**Cheilinus punctatus.**</p> [209, 479, 595.]

Cheilinus punctatus, *Benn. Proc. Comm. Zool. Soc.* i. p. 167; *Günth. Fish.* iv. p. 127.
—— punctulatus, *Cuv. & Val.* xiv. p. 87, pl. 396; *Peters, Wiegm. Arch.* 1855, p. 264.

This species is very closely allied to *C. chlorurus,* Bl., and it may be regarded as the western representative of that fish, which inhabits the East-Indian archipelago.

Colour in life.—Varying from greenish to reddish brown; each scale with several red spots; head and opercles similarly spotted, the dots on the occiput and snout being generally larger and brighter than the others. Sometimes a red or orange spot at the superior angle of the operculum and on the axil of the pectoral. Dorsal marbled with green and dark; it has an orange margin, a green intramarginal line, and some red spots below; the posterior angle is lighter than the remainder of the fin. Anal red, with green margin and intramarginal line, sometimes variegated or banded with green and red. Caudal and ventrals variegated with green and brown.

Zanzibar. Mozambique. Mauritius.

309. <p style="text-align:center">**Cheilinus lunulatus.**</p> [115.]

Labrus lunulatus, *Forsk.* p. 37.
Cheilinus lunulatus, *Rüpp. Atl. Fische,* p. 21, t. 6. f. 1 (bad); *Cuv. & Val.* xiv. p. 88; *Günth. Fish.* iv. p. 130.

Aden. Red Sea.

310. **Cheilinus radiatus.** [210.]

Sparus radiatus, *Bl. Schn.* p. 270, t. 56.

Cheilinus commersonii, *Benn. Proc. Comm. Zool. Soc.* i. p. 167.

—— coccineus, *Rüpp. Atl. Fische*, p. 23.

—— diagramma, *Cuv. & Val.* xiv. p. 98.

—— radiatus, *Bleck. Atl. Ichth.* i. p. 68, t. 26. f. 1; *Günth. Fish.* iv. p. 131.

Zanzibar. From east coast of Africa to the Western Pacific.

311. **Cheilinus rhodochrous,** sp. n. Plate XI. fig. 3. [355, 589.]

D. $\frac{9}{10}$. A. $\frac{3}{8}$. L. lat. 20. L. transv. 2/7.

The height of the body is contained four times and a half, and the length of the head three times and a half in the total length. Snout elongated, conical, the lower jaw slightly the longer. The diameter of the eye is two-fifths the length of the snout. The anterior tubules of the lateral line have a single lateral branch. The second series of cheek-scales does not cover the lower præopercular limb. Caudal rounded, sometimes the upper and lower rays very slightly produced. Ventrals rounded.

Colour in life.—Red, with blue spots and reticulated markings on the snout, opercles, and occiput. An orange streak from lower posterior margin of eye to opercular margin, just above the pectoral. Several blue lines across throat, from the lower series of cheek-scales on one side to that of the other. Spinous portion of dorsal bluish anteriorly, darker between the first and second spines, with a yellow line and darker blue spots near the edge; margin reddish. Anal and soft dorsal red. Caudal red, variegated with darker and with a few white spots, and a broad whitish band across the base.

Length 6 inches.

Zanzibar.

312. **Cheilinus calophthalmus,** sp. n. Plate XI. fig. 4. [517.]

D. $\frac{9}{10}$. A. $\frac{3}{8}$. L. lat. 22. L. transv. 2/7.

The height of the body is contained three times and two-thirds in the total length, and equals the length of the head. Cheek-scales in two series, the lower of which consists of three, and covers the lower præopercular limb. Snout somewhat produced; the length of the præorbital equals the diameter of the eye; the posterior margin of the orbit is rather nearer the end of the operculum than the extremity of the snout. Tubules of the lateral line slightly branched in front. None of the fin-rays produced. Colour dark reddish brown; head and opercles dotted with red; a small black spot on the middle of the interrupted portion of the lateral line, and another at the root of the caudal. A black ocellus, with yellow edges, between the first and second dorsal spines; below this is a bright carmine spot, extending from the first spine to the third.

Length $5\frac{1}{2}$ inches.

Zanzibar.

EPIBULUS, *Cuv.*

313. **Epibulus insidiator.** [258, 258 A, 291.]

Sparus insidiator, *Pall. Spicil. Zool.* viii. p. 41, t. 5. f. 1; *Bl. Schn.* p. 278.
Epibulus insidiator, *Cuv. & Val.* xiv. p. 110, pls. 398, 399; *Cuv. Règne Anim. III. Poiss.* pl. 88; *Bleek.
Atl. Ichth.* i. p. 74, t. 22. f. 3; *Günth. Fish.* iv. p. 135.

$$D. \frac{9}{10\text{-}11}. \quad A. \frac{3}{8\text{-}9}.$$

Var. *a* [258]. Colour brownish green, each scale with a dark-green margin.

Var. *b* [258 A]. Colour as the last, but instead of the green margin each scale has a brown vertical streak or spot on the base.

Var. *c* [291]. Brilliant yellow; this variety has been named by Van Hasselt *Epibulus aureus.*

Zanzibar. *Indian Ocean and Archipelago.*

ANAMPSES, *Cuv.*

314. **Anampses cæruleopunctatus.** [192 A.]

Anampses cæruleopunctatus, *Rüpp. Atl. Fische,* p. 42, t. 10. f. 1; *Cuv. & Val.* xiv. p. 5; *Cuv. Règne Anim. III. Poiss.* pl. 87. f. 2; *Bleek. Atl. Ichth.* i. p. 104, t. 24. f. 2; *Günth. Fish.* iv. p. 135.

Zanzibar. *Mauritius. Red Sea.*

315. **Anampses meleagris.** [356, 634.]

Anampses meleagrides, *Cuv. & Val.* xiv. p. 12; *Bleek. Atl. Ichth.* i. p. 102, t. 24. f. 1.
—— meleagris, *Günth. Fish.* iv. p. 135.

Var. *a* [356]. That usually described.

Var. *b* [634]. General colour of body dark brown; the spots are smaller than in var. *a,* and those on the middle of the body assume the appearance of vertical lines.

Zanzibar. *Mauritius. Amboyna.*

316. **Anampses amboinensis.** [669.]

Anampses amboinensis, *Bleek. Atl. Ichth.* i. p. 103, t. 25. f. 2; *Günth. Fish.* iv. p. 137.

Zanzibar. *Amboyna.*

317. **Anampses diadematus.** [254.]

Anampses diadematus, *Rüpp. N. W. Fische,* p. 21, t. 6. f. 3; *Günth. Fish.* iv. p. 137.
? Anampses lineolatus, *Benn. Proc. Comm. Zool. Soc.* iii. 1837, p. 208.

Zanzibar. Aden. *Mauritius? Red Sea.*

N 2

HEMIGYMNUS, *Günth.*

318.   **Hemigymnus fasciatus.**   [257.]

Mullus fasciatus, *Thunb. Reise nach Japan,* iv. p. 351, t. 314.
Labrus fasciatus, *Bl.* t. 290.
Scarus quinquefasciatus, *Benn. Fishes of Ceylon,* pl. 23.
Tautoga fasciata, *Cuv. & Val.* xiii. p. 203, pl. 379.
Cheilinus blochii, *Cuv. & Val.* xiv. p. 108.
Hemigymnus fasciatus, *Günth. Ann. & Mag. Nat. Hist.* 1861, viii. p. 386, and *Fish.* iv. p. 138; *Bleek. Atl. Ichth.* i. p. 141, t. 46. f. 2.

Zanzibar.   *Mauritius to East-Indian archipelago.*

319.   **Hemigymnus melanopterus.**   [193, 533.]

Labrus melapterus, *Bl.* t. 285.
Tautoga melapterus, *Cuv. & Val.* xiii. p. 311.
Hemigymnus melanopterus, *Günth. Fish.* iv. p. 139.
—— melapterus, *Bleek. Atl. Ichth.* i. t. 45. f. 2, 3.

The specimen marked [533], which is an adult, differs in coloration from those usually observed at Zanzibar. The prevailing colour of the body is bluish green—each scale, excepting those on belly, having a pale-blue crescent, beyond which is a greenish margin; those on the upper part of the body are darkest. Forehead purple, with green lines radiating from the eyes, and similarly coloured spots between them. The middle of the operculum and part of the præoperculum red, with waved blue lines. Snout green. Dorsal brownish, with a green basal, a red marginal, and a blue intramarginal line; between the green and blue lines on the spinous portion is a single series of blue spots; there are several series on the soft portion. Anal similarly coloured; caudal marbled with blue, green, and red; pectorals pale green; ventrals white, marked with blue and red.

Zanzibar.   *East-Indian archipelago. Australia.*

STETHOJULIS, *Günth.*

320.   **Stethojulis strigiventer.**   [285, 412.]

Julis strigiventer, *Benn. Proc. Zool. Soc.* 1832, p. 184; *Cuv. & Val.* xiii. p. 468.
Stethojulis strigiventer, *Günth. Ann. & Mag. Nat. Hist.* 1861, viii. p. 386, and *Fish.* iv. p. 140; *Bleek. Atl. Ichth.* i. p. 135, t. 43. f. 1.

Var. *a* [285]. Head and back above pectorals greenish olive, with darker spots and undulating lines on the fore part; below pectorals whitish, with five or six olive longitudinal lines, on which there are some black dots. A double line, dark olive above, blue below, from snout, touching lower margin of orbit, to angle of operculum; a shorter and less distinct blue line through upper part of orbit. Caudal olive, with about three white cross bands; a black spot below last dorsal rays, two others on the base of the caudal above the lateral line.

Var. *b* [412]. Colour bright green; lower parts of the sides with several pale-green longitudinal lines, on which there are some black dots. A blue, brown-edged band from mouth, below eye, to operculum; sometimes an ocellated spot at the end of the dorsal and on the upper part of the base of the caudal.

Zanzibar. Mauritius. From East Africa to Australia.

321. Stethojulis albovittata. [321.]

Labrus albovittatus, (*Bonnat.*) *Lacép.* iii. pp. 443, 509.
Julis balteata, *Quoy & Gaim. Voy. Uran.* p. 267, pl. 56. f. 1; *Cuv. & Val.* xiii. p. 475.
Stethojulis albovittata, *Günth. Ann. & Mag. Nat. Hist.* 1861, viii. p. 386, and *Fish.* iv. p. 141; *Bleek. Atl. Ichth.* p. 132, t. 44. f. 5.

Colour in life.—Dark olive-green on back, shaded off to white on belly; head emerald-green; the longitudinal stripes are brilliant blue green, becoming deep red in spirits, and white or reddish in a dried state; *never red in life.* Throat purple. Dorsal yellowish, with a white margin; anal, pectorals, and ventrals colourless; caudal rosy.

Zanzibar. Madagascar. East-Indian archipelago. Sandwich Islands.

322. Stethojulis renardi. [351.]

Renard, i. t. 30. f. 160.
Julis renardi, *Bleek. Nat. Tydschr. Ned. Ind.* 1851, *Banda,* i. p. 253.
Stethojulis renardi, *Bleek. Atl. Ichth.* i. p. 132, t. 43. f. 2; *Günth. Fish.* iv. p. 141.

Zanzibar. East-Indian archipelago.

323. Stethojulis interrupta. [593.]

Julis interrupta, *Bleek. Nat. Tydschr. Ned. Ind.* 1851, *Banda,* i. p. 252.
Stethojulis interrupta, *Günth. Ann. & Mag. Nat. Hist.* 1861, viii. p. 386, and *Fish.* iv. p. 142; *Bleek. Atl. Ichth.* i. p. 133, t. 44. f. 4.

Zanzibar. East-Indian archipelago.

324. Stethojulis axillaris. [403.]

Julis axillaris, *Quoy & Gaim. Voy. Uran. Zool.* p. 272; *Cuv. & Val.* xiii. p. 472.
Stethojulis axillaris, *Bleek. Atl. Ichth.* p. 136, t. 44. f. 7; *Günth. Fish.* iv. p. 142.

Colour in life.—Green, the upper part of head and body densely dotted with lighter; on the lower part each scale has a darker centre; a yellowish band from the angle of mouth below eye, as far as the posterior limb of the operculum; a brilliant orange spot behind the angle of operculum; a bluish spot on the base of the pectorals; two or three black blue-edged spots on the base of the tail.

Zanzibar. Johanna. *Madagascar. East-Indian archipelago. Sandwich Islands.*

325.   **Stethojulis kalosoma.**   [411.]

Stethojulis kalosoma, *Bleek. Atl. Ichth.* i. p. 134, t. 43. f. 4; *Günth. Fish.* iv. p. 142.

Colour in life.—Upper part of trunk brownish olive; head with a reddish tinge: throat and belly silvery; a blue line with red margins from the snout, below the eye, to the operculum. Back minutely dotted with faint blue, sides and tail reticulated with the same; a brownish band with blue margins from the angle of the operculum, ceasing above the first anal rays; large brown spots on the lower parts of the sides; a reddish spot with a brown centre above the axil of the pectorals. Dorsal light, with small brown spots; the other fins immaculate.

Zanzibar. *East-Indian archipelago.*

PLATYGLOSSUS, *Klein**.

326.   **Platyglossus dussumieri.**   [353.]

Russell, ii. f. 120.
Julis dussumieri, *Cuv. & Val.* xiii. p. 478, pl. 387; *Cant. Mal. Fish.* p. 236.
Halichœres nigrescens, *Bleek. Atl. Ichth.* i. p. 118, t. 37. f. 4.
Platyglossus dussumieri, *Günth. Fish.* iv. p. 143.

Zanzibar. *East-Indian seas and Archipelago. China.*

327.   **Platyglossus scapularis.**   [240, 354.]

Julis scapularis, *Benn. Proc. Comm. Zool. Soc.* 1831, p. 167.
—— cœruleovittatus, *Rüpp. N. W. Fische*, p. 14, t. 4. f. 1.
—— elegans, (*Kuhl & v. Hass.*) *Cuv. & Val.* xiii. p. 467.
—— leschenaultii, *Cuv. & Val.* xiii. p. 467.
Güntheria cœruleovittata, *Bleek. Atl. Ichth.* i. p. 137, t. 32. f. 2.
Platyglossus scapularis, *Günth. Fish.* iv. p. 146.

Zanzibar. *From east coast of Africa to East-Indian archipelago.*

328.   **Platyglossus hortulanus.**   [246, 352.]

Renard, i. t. 11. f. 71, ii. f. 120.
Labrus hortulanus, *Lacép.* iii. p. 516, pl. 29. f. 2.
—— centiquadrus, (*Commers.*) *Lacép.* iii. p. 493.
Sparus decussatus, *Benn. Fishes of Ceylon*, pl. 14.
Halichœres eximius, *Rüpp. N. W. Fische*, p. 16, t. 5. f. 1.

---

* Undetermined specimens [602, 716].

Julis decussatus, *Cuv. & Val.* xiii. p. 433.
—— hortulanus, *Cuv. & Val.* xiii. p. 430.
Hemitautoga centiquadrus, *Bleek. Atl. Ichth.* i. p. 139, t. 32. f. 3.
Platyglossus hortulanus, *Günth. Fish.* iv. p. 147.

Var. *a* [246]. Corresponds to Bleeker's figure; the general coloration is brownish yellow, each scale having a brown vertical streak, generally formed by two spots joined by a lighter band.

Var. *b* [352]. More nearly allied to Rüppell's figure; general colour green, each scale with a dark centre.

Zanzibar. *Red Sea. From East Africa to Polynesia.*

329.   **Platyglossus opercularis.** Plate XII. fig. 1.   [457.j

Platyglossus opercularis, *Günth. Fish.* iv. p. 148.

The height of the body is contained four times and a third in the total length, that of the head three times and four-fifths. A small posterior canine tooth; caudal rounded; ventrals not produced.

Colour reddish brown, marbled with darker and with black. Head with a reddish-violet darker-edged band running from the snout, through the lower half of the eye, to the operculum; a silvery band immediately below the first band. A black spot behind the orbit, and another, lighter, on the extremity of the operculum. Lower third of the head whitish. Each scale on the upper half of the body has a blackish-brown margin. The lower half of the body is whitish. Base of pectorals silvery white, pectorals transparent; ventrals white. Dorsal fin yellowish, with pairs of oblique, reddish-violet, darker-edged streaks; a small black spot between the first two spines; a large black yellow-edged ocellus between the first three rays, below the margin. Caudal yellowish, with the outer half indistinctly transversely striped; anal grey, with some very indistinct whitish spots along the base.—Length 3 inches.

Island of Johanna. *Feejee Islands. Amboyna.*

330.   **Platyglossus bimaculatus.**   [230.]

Halichœres bimaculatus, *Rüpp. N. W. Fische*, p. 17, t. 5. f. 2.
Julis bimaculatus, *Cuv. & Val.* xiii. p. 493.
Platyglossus bimaculatus, *Günth. Fish.* iv. p. 157.

Zanzibar. *Red Sea. Ceylon.*

PSEUDOJULIS, *Bleek.*

331.   **Pseudojulis argyreogaster,** sp. n.   Plate XII. fig. 2.   [584, 627, 684.]

$$D. \frac{9}{12}. \quad A. \frac{3}{12}. \quad L. lat. 27.$$

Height of the body one-sixth, and length of the head one-fourth of the total length.

Colour in spirits.—Uniform light yellowish olive; one specimen with an elongate ovate blackish spot before and below the bend of the lateral line; this spot is absent in another specimen, and represented in a third by a minute dot only; the two last-mentioned specimens have a broad pearl-coloured band along each side of the belly, commencing at the base of the pectorals.

Colour in life.—Green, a violet line from middle of snout (where it meets the corresponding one on the other side) skirting the upper margin of the orbit, nearly as far as the end of the operculum. A similar line on the checks below the orbit, from the angle of the mouth nearly to the præopercular margin. A large violet ring on the upper part of the operculum. Blotch on the sides as before described. Dorsal and anal orange, with a greenish tint and a narrow violet margin.—Length 3½ inches.
Zanzibar.

NOVACULA, *Cuv. & Val.*[*]

332.   **Novacula tæniurus.**   [214, 256, 560.]

Labrus tæniurus, *Lacép.* iii. pp. 448, 518, pl. 29. f. 1.
Xyrichthys tæniurus, *Cuv. & Val.* xiv. p. 54, pl. 393 (not *Bleek.*).
Novacula tæniurus, *Günth. Fish.* iv. p. 172.

Var. *a* [214]. *Colour in life.*—Brown, each scale with an olive-green centre; belly-scales red, with white margins; head immaculate. Dorsal greenish, with oblique brown stripes and spots; anal brownish green, with oblong brown stripes and spots along the rays, and round ones between them; two deep-blue-black spots between the first and second and second and third dorsal spines, each with a yellow margin anteriorly. A broad yellow band across the base of the caudal. An orange spot on axil of pectoral, and a black crescent behind that fin, which is black below and yellow above.—Length 12 inches.

Var. *b* [256, 560]. Similar in general coloration to var. *a*, but without any markings on the dorsal and anal fins except the two anterior spots. The band across the base of the caudal is darker than in var. *a*, and the belly-scales are not red.—Length 13 inches.
Zanzibar. *Mauritius.*

333.   **Novacula macrolepidota.**   [302, 688.]

Labrus macrolepidotus, *Bl.* t. 284.
Julis trimaculatus, *Cuv. & Val.* xiii. p. 386.
Xyrichthys macrolepidotus, *Cuv. & Val.* xiv. p. 59; *Peters, Wiegm. Arch.* 1855, p. 264.
Novaculichthys macrolepidotus, *Bleek. Atl. Ichth.* i. p. 144, t. 31, f. 6.
Novacula macrolepidota, *Günth. Fish.* iv. p. 174.

* Undetermined specimen [578].

The colour in life is bright green; in addition to the three bands radiating from the orbit (which have yellow edges) there are three others, much lighter,—the first from the anterior of the orbit to the maxillary, the second from below the eye, across the chin, and the third across the præopercularum and throat. Dorsal and anal green, with several oblique series of red spots, sometimes confluent.

Zanzibar. *Mozambique. East-Indian archipelago. New Guinea. Waigiou.*

334. **Novacula pavo.** [683.]

Xyrichthys pavo, *Cuv. & Val.* xiv. p. 61, pl. 394; *Bleek. Atl. Ichth.* i. p. 149, t. 29. f. 2.
—— pavoninus, *Cuv. & Val.* xiv. p. 63.
Novacula pavo, *Bleek. Nat. Tydschr. Ned. Ind.* 1856, *Ternate,* vii. p. 379; *Günth. Fish.* iv. p. 175.

Colour in life.—Greenish olive, with three bluish cross bands; a large bright patch on the belly, behind the tips of the pectorals, in which each scale has a bright-blue centre; in other points the colour is as described in Günther's Catalogue of Fishes.

Zanzibar. *Réunion. Mauritius. East-Indian archipelago. Sandwich Islands.*

335. **Novacula aneitensis.** [227.]

Novacula aneitensis, *Günth. Fish.* iv. p. 176.

Zanzibar. *New Hebrides.*

336. **Novacula punctulata.** [441.]

Novacula punctulata, *Cuv. & Val.* xiv. p. 73; *Günth. Fish.* iv. p. 177.

Colour in life.—Pale greenish olive; a large reddish-brown spot on the lateral line below the last dorsal spines, underneath which is a still larger white and blue patch; a pale-blue line from above angle of operculum to behind operculum; another, edged with yellow, from anterior margin of orbit to angle of mouth; upper edge of head bluish. Dorsal orange, with oblique pale-violet wavy lines; anal pale greenish olive, with oblique bright-yellow lines; caudal rosy, with six or seven transverse series of darker spots.

Zanzibar. *China.*

JULIS, *Cuv. & Val.**

337. **Julis lunaris.** [80.]

Labrus lunaris, *L., Bleek. Atl. Ichth.* i. p. 90, t. 33. f. 5; *Günth. Fish.* iv. p. 180, which see for the rest of the synonymy.

Aden. Zanzibar. Seychelles. *From the east coast of Africa to China and Polynesia.*

* Undetermined specimen [402].

338. <div align="center">**Julis hebraica.**</div> [197.]

Labrus hebraicus, *Lacép.* pp. 455, 526.
Julis cingulata, *Quoy & Gaim. Voy. Astrol.* p. 711, pl. 15. f. 3.
—— genivittata, *Cuv. & Val.* xiii. p. 416.
—— hebraicus, *Cuv. & Val.* xiii. p. 423.
—— hebraica, *Günth. Fish.* iv. pp. 186, 508.

Colour in life.—Bluish on the shoulders, the remainder of the fish yellowish green; a bright-yellow band across shoulders, passing below the pectorals, nearly to the edge of the belly; under lip blue; several brilliant blue bands on the head and throat; a deep-blue band, becoming black in a dried state, across the base of the tail. Vertical fins olive-green, a black spot anteriorly on the dorsal; pectorals transparent; ventrals blue.

Zanzibar. *Red Sea. Mozambique. Port Natal. Mauritius. Madagascar.*

339. <div align="center">**Julis trilobata.**</div> [416.]

Labrus trilobatus, *Lacép.* iii. pp. 454, 526 (not *Shaw*).
Julis bicatenatus, *Benn. Proc. Comm. Zool. Soc.* i. 1831, p. 167.
—— quadricolor, *Less. Voy. Coqu. Zool.* ii. p. 139, pl. 35. f. 1; *Cuv. & Val.* xiii. p. 443; *Bleek. Atl. Ichth.* i. p. 93, t. 3 l. f. 3.
—— semiceruleus, *Rüpp. N. W. Fische,* p. 10, pl. 2. f. 1; *Cuv. & Val.* xiii. p. 442.
Scarus georgii, *Benn. Fishes of Ceylon,* pl. 24.
Julis trilobatus, *Cuv. & Val.* xiii. p. 437.
—— formosus, *Cuv. & Val.* xiii. p. 439.
—— æruginosus, *Cuv. & Val.* xiii. p. 441.
—— cyanogaster, *Cuv. & Val.* xiii. p. 444.
—— erythrogaster, *Cuv. & Val.* xiii. p. 447.
—— trilobata, *Günth. Fish.* iv. p. 187.

Zanzibar. *From the east coast of Africa to China and Polynesia.*
The specimens from Zanzibar correspond to var. β of Günther.

340. <div align="center">**Julis dorsalis.**</div> [202.]

Sparus hardwickii, *Benn. Fishes of Ceylon,* pl. 12 (*not* Julis hardwickii, *Gray*).
Julis dorsalis, *Quoy & Gaim. Voy. Astrol. Zool.* iii. *Poiss.* p. 713, pl. 15. f. 5; *Cuv. & Val.* xiii. p. 449;
Bleek. Atl. Ichth. i. p. 94, t. 34. f. 4; *Günth. Fish.* iv. p. 190.

Zanzibar. *From the east coast of Africa to China and Polynesia.*

GOMPHOSUS, *Lacép.*

341. <div align="center">**Gomphosus cæruleus.**</div> [185.]

Gomphosus cæruleus, *Lacép.* iii. p. 101, pl. 5. f. 1, pl. 6. f. 1; *Cuv. & Val.* xiv. p. 29; *Bleek. Atl. Ichth.* i. p. 86, t. 21. f. 5; *Günth. Fish.* iv. p. 192.
—— viridis, *Benn. Fishes of Ceylon,* pl. 30.

Aden. Zanzibar. *East Indies.*

342. Gomphosus varius. [572.]

Gomphosus varius, *Lacép.* iii. p. 104, pl. 5. f. 2; *Günth. Fish.* iv. p. 193.
—— fuscus, *Benn. Fishes of Ceylon*, pl. 3; *Cuv. & Val.* xiv. p. 25.

Zanzibar. Island of Angazilia or Great Comoro. *Mauritius. Polynesia.*

CHEILIO, *Commers.*

343. Cheilio inermis. [215, 221, 249, 290, 434, 442.]

For synonymy see *Günth. Fish.* iv. p. 194.

This fish exhibits great diversity of coloration; the specimens from Zanzibar are as follows:—

a [249]. Body olive-green above, pale blue below; each scale with a brown centre; the lateral band does not extend beyond the opercles.

b [221]. Olive-green above, bright green below; each scale with a light centre; no lateral band; a large red blotch on side below the middle of spinous dorsal.

c [290, 442]. Olive-green above, white below; scales with darker margins; lateral band formed of black spots more or less confluent.

d [215]. Green, with a distinct lateral band.

e [434]. Bright green, each scale with a blue centre; a yellow lateral band, with a series of black spots above it.

Aden. Zanzibar. *From the eastern coasts of Africa to the Western Pacific.*

CORIS, *Gthr.*[*]

344. Coris cuvieri. [282, 588.]

Julis cuvieri, *Benn. Proc. Comm. Zool. Soc.* i. p. 128.
—— stellatus, *Cuv. & Val.* xiii. p. 499.
Coris cuvieri, *Günth. Fish.* iv. p. 199.

Colour in life.—Purple, each scale with a blue or green spot; cheeks green; three brown and four green bands on head; of the former the first is distinct, and proceeds from posterior part of orbit towards upper part of operculum; the second proceeds in a curve from snout towards middle of operculum, and the third from angle of mouth through angle of preoperculum across interoperculum. The first green band is from upper part of orbit to first dorsal spine; the second is between the first and second brown band; the third between the second and third brown band; and the fourth from mandible to base of ventral. Lower part of dorsal and anal green, with irregular oblique carmine streaks; the upper half of the former is brownish violet, with blue marginal, median, and basal lines; the outer half of the anal is similar, but with two blue median lines. The basal portion of the caudal is carmine, with blue spots; beyond this there

* Undetermined specimen [641].

is a greenish, then a blackish portion, also a white margin and black intramarginal line. Inner surface of pectorals violet; outer surface green. Ventrals longitudinally streaked with brown, blue, green, and violet.

Zanzibar. *Mauritius. New Hebrides.*

345. **Coris formosa.** [255.]

Labrus formosus, *Benn. Fishes of Ceylon*, pl. 16.
Coris formosa, *Günth. Fish.* iv. p. 201.

$$ \text{D.} \frac{9}{12}. \quad \text{A.} \frac{3}{12}. $$

The height of the body is contained four times, and the length of the head four times and a half in the total length. Anterior dorsal spines elevated; caudal slightly rounded; ventrals pointed, reaching the vent. Two posterior canine teeth.

Colour in life.—Snout orange, the rest of the body brown, with numerous large round black spots, each nearly as large as the pupil. Two brilliant blue oblique stripes on the head,—the first from a little before the anterior dorsal spine, passing along the superior margins of the orbit towards the upper lip; the second from the second dorsal spine, skirting the lower part of the orbit, to behind the angle of the mouth : the second is the broader of the two. A blue spot on each side of the symphysis of the lower jaw; a green crescent behind axil of the pectoral. Dorsal fin blackish, covered with faint blue, pink-edged spots; it has also a broad reddish margin, having a narrow blackish superior, inferior, and median line. Anal blackish, with a marginal line and an intramarginal series of spots, followed by a second line, all blue. Caudal crimson, with a basal series of black spots, a broad white margin, and a narrow black intramarginal line; ventrals brownish, with a narrow blue interior margin; pectorals reddish at the base, yellowish beyond.

This species has been observed by Colonel Playfair at Zanzibar for the first time since its discovery by Bennett. The specimen agrees as well with Bennett's figure as if the latter had been taken from the former; and Dr. Günther's opinion as to the distinctness of *C. formosa,* Bleek., and *C. formosa,* Benn., is fully confirmed.

Length 15 inches.

Zanzibar. *Ceylon.*

346. **Coris annulata.** [222, 294, 357, 664.]

Labrus annulatus, *Lacép.* iii. pp. 455, 526, pl. 28. f. 3.
Hologymnosus fasciatus, *Lacép.* iii. p. 557, pl. 1. f. 3 ; *Bleek. Atl. Ichth.* i. p. 96, t. 20.
Julis rosea, *Quoy & Gaim. Voy. Astrol.* iii. p. 709, *Poiss.* pl. 15. f. 1.
—— doliatus, *Cuv. & Val.* xiii. p. 50 ٨.
—— annulatus, *Cuv. & Val.* xiii. p. 501, pl. 388.
Coris annulata, *Günth. Fish.* iv. p. 202.

Var. *a* [294, 357]. Var. *b* [222, 664].

Zanzibar. *Indian Ocean.*

347. Coris caudimacula. [281.]

Julis caudimacula, *Quoy & Gaim. Voy. Astrol.* iii. p. 710, *Poiss.* pl. 15. f. 2 ; *Cuv. & Val.* xiii. p. 426.
Hemicoris caudimacula, *Bleek. Atl. Ichth.* i. p. 106, t. 36. f. 2.
Coris caudimacula, *Günth. Fish.* iv. p. 205.

Colour in life.—Greenish, with four longitudinal lines, which are blue on snout, red on head and shoulders, and orange towards tail ; all with blue edges. There is another from the chin to the root of the ventrals, and one from the angle of the operculum, descending obliquely behind the pectorals to the anal. A blue spot on the angle of the operculum, and a black blotch on the root of the caudal. Dorsal orange, with a series of green blue-edged spots along the base, and two or three series of light-blue spots above. Anal orange, with reddish base enclosing orange spots ; four rows of blue spots, and stripes beyond. Caudal with a blue semicircular band from the middle of one lobe to that of the other, cutting off, as it were, the tips of each lobe ; within this there is a broad orange band of similar shape, enclosing bluish spots.

Zanzibar. *Mozambique. Mauritius.*

348. Coris frerei, sp. n. Plate XIII. [624.]

D. $\frac{9}{12}$. A. $\frac{3}{12}$. L. lat. 86.

A posterior canine tooth. The height of the body is contained three times and a half, and the length of the head to the extremity of the opercular lobe four times in the total length without caudal. The two anterior dorsal spines produced, the first being as long as the head ; the third and fourth are shortest ; thence they increase in length posteriorly. The middle dorsal rays are the longest, and half as long as the head. Caudal truncate. Ventrals narrow, long, but rather shorter than the head.

Colour of head and neck fawn, of body bluish violet ; each scale with a round blue or green spot. There are about nine dark-blue cross bands, the first crossing the body above the vent, and each being about as broad as the interspace which separates it from its fellow. Along the median line of the forehead there is a green band, which becomes sinuous in front of the eye, and throws out branches to each side ; there is another from the second dorsal spine to the upper margin of the orbit ; thence it runs at an angle to the middle of the maxillary, and again takes a curve to the angle of the mouth ; a third, of horseshoe-shape, commences on the root of the ventral, runs forward to and along the lower margin of the interoperculum, thence forward to the joint of the mandible, ascends in a curve to the lower margin of the orbit, runs backwards across the suprascapular region, and is gradually lost on the anterior portion of the lateral line ; a fourth skirts the opercular and subopercular margins. These bands are brilliant green, with purple edges, becoming red in a dried state. There are several similarly coloured round spots on the head and opercles. The lower two-thirds of the dorsal is violet, densely covered with green and blue spots, sometimes confluent into short lines ; next follows a reddish band, and beyond this a reddish-brown margin, with four or five brilliant cobalt lines or series of spots. The anal has a broad brown margin, with an upper and lower blue line and several longitudinal series of spots between them ; below this the fin is green, shaded off into brownish violet at the base; the latter colour is spotted with green. Caudal brownish violet, with blue ocelli and a red margin.

Length 20 inches.

Zanzibar.

We have named this fish after Sir H. E. Bartle Frere, G.C.S.I., Governor of Bombay, whose assistance towards the publication of this work we gratefully acknowledge.

CYMOLUTES, Gthr.

349.   **Cymolutes prætextatus.**   [250, 440.]

Julis prætextata, *Quoy & Gaim. Voy. Astrol. Poiss.* p. 112, pl. 15. f. 4 ; *Cuv. & Val.* xiii. p. 508.
Xyrichthys torquatus, *Cuv. & Val.* xiv. p. 54, pl. 392.
—— novaculoides, *Bleek. Nat. Tydschr. Ned. Ind.* 1853, Amboyna, iii. p. 122.
Novacula xyrichthyoides, *Bleek. Nat. Tydschr. Ned. Ind.* x. p. 488.
Cymolutes prætextatus, *Günth. Fish.* iv. p. 207 ; *Bleek. Atl. Ichth.* p. 146, t. 31. f. 1.

Varietas nova.—Colour pale olive; head immaculate; body covered with narrow brown lines angularly bent, the angle pointing backwards; these are generally darker on the shoulder. Sometimes a round brown spot below the lateral line and under the sixth dorsal spine; out of four specimens this spot was absent in two. Dorsal pale violet, with irregular reddish lines and spots; the specimens without the lateral blotch have a deep-black margin to the spinous portion. Anal uniform orange or rosy; caudal orange, with yellowish transverse lines or spots. Ventrals and pectorals transparent.

Zanzibar. *Mauritius. East-Indian archipelago.*

PSEUDODAX, Bleek.

350.   **Pseudodax moluccensis.**   [449.]

Odax moluccanus, *Cuv. & Val.* xiv. p. 305, pl. 408.
Pseudodax moluccanus, *Bleek. Atl. Ichth.* i. p. 80, t. 22. f. 2, t. 18. f. 5.
—— moluccensis, *Günth. Fish.* iv. p. 208.

Zanzibar. *East-Indian archipelago.*

SCARICHTHYS, Bleek.*

351.   **Scarichthys auritus.**   [247.]

Scarus auritus, (*Kuhl & v. Hass.*) *Cuv. & Val.* xiv. p. 218.
—— vaigiensis, *Cuv. & Val.* xiv. p. 214.

* Undetermined specimen [466].

Scarus nævius, *Cuv. & Val.* xiv. p. 253.
Callyodon chlorolepis, *Richards. Voy. Sulph. Ichth.* p. 137, pl. 64. figs. 4–7.
Scarichthys auritus, *Bleek. Atl. Ichth.* i. p. 15, t. 1. f. 3; *Günth. Fish.* iv. p. 213.
Zanzibar. *From Seychelles to New Hebrides.*

352. **Scarichthys cæruleopunctatus.** [601.]
Scarus cæruleopunctatus, *Rüpp. N. W. Fische,* p. 24, t. 7. f. 3; *Cuv. & Val.* xiv. p. 262.
—— bottæ, *Cuv. & Val.* xiv. p. 262.
Scarichthys cæruleopunctatus, *Bleek. Atl. Ichth.* i. p. 16, t. 1. f. 2; *Günth. Fish.* iv. p. 213.

In Günther's Catalogue of Fishes the formula of the anal fin is given as $\frac{2}{8}$, and it is stated as a generic character that the scales of the back cover the base of the dorsal fin. In the specimen from Zanzibar the anal fin is $\frac{3}{6}$, and the dorsal fin is entirely free from scales, in which respects it agrees with Bleeker's description.
Zanzibar. *Red Sea. East-Indian archipelago.*

CALLYODON, *Gron.*

353. **Callyodon viridescens.** [434 A, 544, 604.]
Callyodon viridescens, *Rüpp. N. W. Fische,* p. 23, t. 7. f. 2; *Günth. Fish.* iv. p. 214.

This species is distinguished by having constantly a black base to the pectoral and a more or less distinct black spot between the first and third dorsal spines. None of these characteristics are visible in *C. genistriatus*; in other respects the specimens vary much in colour.

Var.-*a.* Brown, marbled and spotted with darker.

Var. *b.* Olive; belly white, nearly every scale with a brick-red centre; the whole body densely covered with rather irregular indistinct white points. Head, opercles, and middle of trunk with a number of scattered red spots, each having a dark centre. A rather indistinct light lateral band from the angle of the operculum to the root of the caudal. Two red lines pass through the eye at the upper posterior margin, and go round the lower lip; the posterior one touches the angle of the mouth; sometimes a third and shorter line, parallel to the above, below the orbit. Dorsal and caudal indistinctly spotted with reddish yellow; anal with a blackish margin and two longitudinal reddish bands anastomosing with each other and enclosing whitish round spots; ventrals white; pectorals yellow. Black spots on dorsal and pectorals as before described.

Length 3 to 5 inches.
Zanzibar. *Red Sea.*

354. **Callyodon genistriatus.** [542, 582.]
Callyodon genistriatus, *Cuv. & Val.* xiv. p. 293; *Bleek. Atl. Ichth.* i. p. 13, t. 1. f. 1; *Günth. Fish.* iv. p. 215.

Colour of adult in life.—Greenish red, each scale with a vertical red mark at the base, where the next scale overlaps. Head with red lines radiating from the eye, and several red marks on the chin, opercles, and forehead. Dorsal green, with orange margins and a similarly coloured line above the top of the scaly sheath, becoming interrupted and forming spots towards the end. Anal green, with orange marginal and median lines. Caudal green, with a white margin and about four vertical orange bars between the lobes.—Length 10 inches.

There are two other specimens in this collection, 4 and 4½ inches long [582], which are brownish, spotted all over with whitish; fins variegated with lighter and darker brown, and a row of red spots along the base of the anal rays. These we consider to be the young of *C. genistriatus*. They are very similar to the nominal species figured by Bleeker as *C. spinidens*. It appears that the small teeth within the front series of the upper jaw is a characteristic developed in a varying degree in young specimens, which probably disappears entirely with age; the rounded caudal of immature specimens becomes emarginate in adults; we have therefore little doubt that *C. spinidens* is founded on young specimens of *C. genistriatus*.

Zanzibar. East-Indian archipelago.

PSEUDOSCARUS, *Bleek.**†

355.   **Pseudoscarus harid.**   [300, 607.]

Scarus harid, *Forsk.* p. 30; *Rüpp. Atl. Fische,* p. 80, t. 21. f. 1.
—— mastax, *Rüpp. Atl. Fische,* p. 80, t. 21. f. 2, and *N. W. Fische,* p. 28; *Cuv. & Val.* xiv. p. 246.
Pseudoscarus mastax, *Bleek. Atl. Ichth.* i. p. 35, t. 10. f. 1.
—— harid, *Günth. Fish.* iv. p. 220.

Zanzibar. Red Sea. Java. Cocos.

356.   **Pseudoscarus javanicus.**   [537, 673.]

Pseudoscarus javanicus, *Bleek. Atl. Ichth.* i. p. 36, t. 11. f. 3; *Günth. Fish.* iv. p. 220.

Colour in life.—Greenish; each scale with a vertical reddish mark on the base. Lips with an orange band between two green ones; the exterior green band of the upper lip takes a bend at the corner of the mouth and proceeds to the eye, skirting the lower margin. A green band from symphysis of lower jaw to anal. Dorsal and anal violet, with green margins and broad basal bands. Caudal green, with a violet band along the middle of each lobe. Pectorals green, with a violet centre.

Zanzibar. Java.

---

* Although in many instances we have given the colours in life of the species of this genus, these must not be implicitly relied on as specific characters. They vary exceedingly with age, sex, and season; we would particularly direct attention to *P. troschelii*, Bleek., as an instance in point.

† Undetermined specimens [30, 126, 519].

357. **Pseudoscarus maculosus.** [12, 545.]

Scarus maculosus, *Lacép.* iv. pp. 5, 21, pl. 1. f. 3; *Cuv. & Val.* xiv. p. 235.
—— guttatus, *Bl. Schn.* p. 294.
Pseudoscarus maculosus, *Günth. Fish.* iv. p. 223.

Colour in life.—Yellowish above, rosy on the belly, the scales having blue centres, some of which are more strongly marked than others, and form about five interrupted cross bands. A blue streak above and passing through upper part of eye; another from posterior part of orbit; a third below it, skirting the inferior margin, and extending to the angle of the mouth; a blue band round upper lip; another at some distance beneath the lower one, and several blue markings on the opercles. Dorsal rosy, with a blue margin and a blue basal and median series of spots. Anal similar, without the median series of spots. Caudal rosy, with a blue exterior margin to each lobe and a few blue spots between them. Ventrals rosy, with blue anterior margins. Pectorals also with a blue superior margin.

The *Pseudoscarus pyrrhostethus* of Bleeker approximates so closely to the *P. maculosus* of this collection, that we are almost inclined to consider them the same species.

Aden. Zanzibar. Ibo. *Mozambique. Mauritius.*

358. **Pseudoscarus globiceps.** [288.]

Scarus globiceps, *Cuv. & Val.* xiv. p. 179; *Jen. Voy. Beagle, Fishes,* p. 106.
Pseudoscarus globiceps, *Günth. Fish.* iv. p. 224.

Zanzibar. *Tahiti.*

359. **Pseudoscarus nigripinnis,** sp. n. Plate XV. fig. 2. [438, 670.]

$$D. \frac{9}{10}. \quad A. \frac{3}{9}.$$

Jaws whitish, strong, with the denticulation very perceptible, not only on the margin, but on the whole surface of the bone; no separate conical teeth; lip broad, covering more than half of the upper jaw. Scales of the cheek in three series; the middle one consisting of six, and the lower one of two or three scales; the lower præopercular limb is entirely naked. The anterior dorsal spines are highest. Fourteen pectoral rays. Caudal in young specimens slightly rounded; in older examples it is truncated, with the lobes very slightly produced.

Colour yellowish olive; each scale with a darker base and margin, those on the lower half of the fish being nearly black. The scales of the thoracic region have a small black spot in the centre. A dark stripe across the forehead between the anterior margins of the orbits. Another similar stripe from the lower anterior margin of the orbits, passing round the throat just behind the angle of the mouth. The front part of the dorsal, namely the membrane between the first two spines and the base of the following one, yellowish, as is also the margin of the posterior angle; the remainder of the

P

fin deep black. Anal blackish; ventrals blackish, with yellow margins; pectorals yellow; caudal blackish, sometimes with a white posterior margin.

Zanzibar.

Length 12 inches.

It is just possible that the description given by M. Valenciennes of his *Scarus lunulatus* may be intended for this species; but the original specimen appears to be lost, and the description is insufficient for identification.

360.   **Pseudoscarus capitaneus.**   [507.]

Cacatoe-capitano, *Renard*, pl. 20. f. 112 (very bad).
Scarus capitanus, *Cuv. & Val.* xiv. p. 228, pl. 403.
Pseudoscarus capitaneus, *Günth. Fish.* iv. p. 228.

Colour in life.—Brownish, with three rows of large whitish spots,—the first, consisting of three spots, near the base of dorsal; the second below the lateral line, consisting of about six; and the third, also of about six, from axil of pectoral. Caudal, anal, and dorsal uniform brown.

Zanzibar. *Mauritius. Mozambique.*

361.   **Pseudoscarus macrochilus.**   [273.]

Pseudoscarus macrocheilos, *Bleek. Atl. Ichth.* i. p. 38, t. 15. f. 1.
—— macrochilus, *Günth. Fish.* iv. p. 228.

Colour in life.—Greenish grey, each scale with a lighter centre; dorsal blackish, the posterior rays with a reddish tint; caudal and anal red; ventrals and pectorals grey.

Zanzibar. *Java.*

362.   **Pseudoscarus rubro-violaceus.**   [547.]

Pseudoscarus rubro-violaceus, *Bleek. Atl. Ichth.* i. p. 37, t. 13. f. 3; *Günth. Fish.* iv. p. 229.

Zanzibar. *Java.*

363.   **Pseudoscarus bataviensis.**   [189, 536.]

Pseudoscarus bataviensis, *Bleek. Atl. Ichth.* i. p. 48, t. 12. f. 3; *Günth. Fish.* iv. p. 231.

The specimens from Zanzibar were found to differ somewhat from those previously described; there are two or three conical teeth at the inner angle of the lower jaw, which are stronger than those on the upper. Having examined the typical specimen of the species ceded by Dr. Bleeker to the British Museum, we find that the teeth in the lower jaw are absent on one side only, but in the other they are as well developed as in the Zanzibar specimen. In preserved specimens a large light-yellowish blotch on the lower part of operculum (green in life) is very distinct, and appears to be characteristic of the species.

The formula of the anal fin is $\frac{3}{5}$, and in some specimens the posterior half of the fish is of a lighter colour than the anterior.

Zanzibar. *Batavia.*

364. Pseudoscarus pentazona. [550.]

Pseudoscarus pentazona, *Bleek. Atl. Ichth.* i. p. 46, t. 11. f. 1; *Günth. Fish.* iv. p. 231.

D. $\frac{9}{6}$. A. $\frac{3}{5}$.

Colour in life.—Pale grey, with about five rather indistinct darker cross bands, each scale with a reddish vertical line at the base. Lips blue; a separate blue line parallel to each; between these the colour is rosy; the upper line is continued to the middle of the anterior part of orbit. A blue line along the lower edge of the fish from symphysis of jaws to vent. Dorsal and anal brownish, clouded with purple in the centre, and with purple marginal lines; caudal purplish brown, with darker margins to each lobe.

Zanzibar. *Celebes.*

365. Pseudoscarus erythrodon. [548.]

Scarus erythrodon, *Cuv. & Val.* xiv. p. 255.

Pseudoscarus sumbawensis, *Bleek. Atl. Ichth.* i. p. 47, t. 15. f. 4; *Günth. Fish.* iv. p. 232.

Colour in life.—Blackish brown, each scale with a dark brown margin; tail lighter than body, a violet tint on thoracic region; snout yellowish brown; dorsal and anal brownish, with a narrow black marginal line; ventrals brown; pectorals transparent brown.

Zanzibar. ? *Mauritius. East-Indian archipelago.*

366. Pseudoscarus nuchipunctatus. [213, 541.]

Scarus nuchipunctatus, *Cuv. & Val.* xiv. p. 224.

—— limbatus, *Cuv. & Val.* xiv. p. 271.

Pseudoscarus nuchipunctatus, *Bleek. Atl. Ichth.* i. p. 31, t. 10. f. 2; *Günth. Fish.* iv. p. 233.

D. $\frac{9}{10}$. A. $\frac{2-3}{9}$.

Spec. *a* [213]. *Colour in life.*—Head and belly deep bright red; the remainder reddish brown, with lighter longitudinal stripes, which are of the same breadth as the interspaces between them; these occur in the middle and on the top and bottom of each scale. A deep-greenish, almost black stripe passes through the upper part of the orbit across the operculum; another round the upper lip; a third, more or less interrupted, passes beneath the orbit, round the lower lip; and a fourth, parallel to the last, lower down. There is a similarly coloured longitudinal line along the junction of the branchiostegous membranes, and another on each side along the inferior margin of the præoperculum. Dorsal and anal reddish, with blue or green marginal lines. Pectorals and ventrals reddish. Caudal with a narrow posterior blue margin.

Length 8½ inches.

Spec. *b* [541]. Corresponds to the description and figure in Bleeker's 'Atlas Ichthyologique.'
Length 12 inches.
Zanzibar. *Indian Ocean and Archipelago.*

367. **Pseudoscarus cyanognathus.** [523, 546, 662.]

Pseudoscarus cyanognathus, *Bleek. Atl. Ichth.* i. p. 32, t. 11. f. 2; *Günth. Fish.* iv. p. 234.

$$D. \frac{9}{10}. \quad A. \frac{2-3}{9}.$$

Colour in life. Spec. *a* [546].—Green, each scale with a reddish base. Four longitudinal stripes on belly, the lowest from symphysis of lower jaw to first anal spine. Lips blue, coalescing at the angle of the mouth into a single green line, which skirts the lower margin of the orbit, and extends to the posterior limb of the præoperculum. A second blue line parallel to the lower lip; a blue streak along the lower limb of the præoperculum. Vertical fins yellowish, with green marginal, median, and basal bands, Caudal green, with a double bluish margin to the lobes, between which the colour is reddish.—Length 10½ inches.

Spec. *b* [523]. Similar in colour to *a*. The longitudinal lines on the belly are hardly visible; the green line below the eye, which in *a* was a narrow streak, has become a broad band, and extends to the posterior of operculum; the margin of the anal fin is very broad, occupying about one-half of its breadth.— Length 11 inches.

Spec. *c* [662]. The band below the eye has widened into a broad green patch over the opercles. The scales on the posterior part of the tail and the sheath of the dorsal are uniform bright green, without reddish centres. The longitudinal lines on the belly are inconspicuous.—Length 13 inches.
Zanzibar. *Java.*

368. **Pseudoscarus caudofasciatus.** [450, 534.]

Pseudoscarus caudofasciatus, *Günth. Fish.* iv. p. 238,

$$D. \frac{9}{10}. \quad A. \frac{3}{9}.$$

Colour in life.—Bluish grey, with three or four transverse yellow bands, and an indistinct longitudinal one from angle of mouth to root of pectoral. Vertical fins red, immaculate; spinous portion of dorsal with a very narrow blackish margin. Pectorals and ventrals white.
Zanzibar. *Mauritius.*

369. **Pseudoscarus troscheli.** Plate XIV. figs. 1, 2, 3. [509, 522, 512.]

Pseudoscarus troscheli, *Bleek. Atl. Ichth.* i. p. 25, t. 7. f. 2; *Günth. Fish.* iv. p. 237.

$$D. \frac{9}{10}. \quad A. \frac{3}{9}.$$

The examples of this species differ widely in coloration; but it may be easily recognized by the powerfully developed jaws, which have the edge distinctly denti-culated, the small eye, and the peculiar shape of the præoperculum.

The following is the colour in life of three distinct varieties:—

Var. *a* [509]. Jaws green. Caudal truncated. Body green, each scale with a purple base, except on the extremity of the tail, which is entirely green. Snout purple; upper lip with a red margin next the jaw; throat blue: a red margin to lower lip, extending as far as the cheek-scales; a red spot at the symphysis of the lower jaw, and a red line from the first scale of the lower cheek-series across throat and down margin of interoperculum; one or two red spots on interoperculum. Dorsal green, with a yellow median band and a blue margin. Anal green, with a purple base and a blue margin. Pectorals green; a red streak along the second ray. Caudal green, with a blue margin.—Length 10 inches.

Var. *b* [522]. Jaws rosy, with two conical teeth at the angle of the upper jaw. The caudal lobes slightly produced. Colour purple brown, each scale with a darker basal membrane, except on the extremity of the tail, which, like var. *a*, is lighter than the rest of the body. Snout rich deep brown. Dorsal brown, with a black median band and marginal line. Anal and ventral darker. Pectorals yellow, with a brown base.—Length 10 inches.

Var. *c* [512]. Jaws grey. Caudal lobes somewhat produced. Body rose-coloured, lips green; dorsal reddish, with a green marginal line. Anal green, with a red margin and a green marginal line. Caudal with narrow upper and lower green margins and a broad green margin posteriorly.—Length 13 inches.

Zanzibar. *Java.*

370. Pseudoscarus chloromelas, sp. n. Plate XV. fig. 1. [617.]

$$D. \frac{9}{10}. \quad A. \frac{3}{5}.$$

Jaws grey, very strong, distinctly denticulated on the margins, and with one or two conical teeth at the posterior angle of the upper jaw. Lips rather narrow, double only posteriorly. Forehead swollen. Two series of scales on the cheek, each of which contains five scales, the præopercular limb being entirely naked. Caudal slightly convex. Dorsal spines subequal. Fifteen pectoral rays.

The naked portions of the head deep blue, the general colour of the rest of the head and body black: the body-scales are irregularly marked with opaque yellow green; some have one or more spots, others irregularly curved lines, others annuli, while others have a large spot in the centre. Dorsal blue black, with a pale blue margin, broadest in front, and a series of irregular green spots along the base and middle. Anal greenish, with a blue margin, a brown intramarginal and basal band, with a series of similarly coloured spots between them. Ventrals and pectorals blue

black, the former with a pale blue exterior margin. Caudal blue black, each lobe with
a narrow pale blue margin.

Length 12½ inches.

Zanzibar.

Family GERRIDÆ.

GERRES, *Cuv.*

371.   **Gerres acinaces.** Plate XVI. fig. 1.   [46.]

Gerres acinaces, Bleek. Nat. Tydschr. Nederl. Ind. 1854, vi. p. 194; *Günth. Fish.* iv. p. 262.

D. $\frac{9}{10}$. A. $\frac{3}{7}$. L. lat. 45. L. transv. $\frac{6}{11}$.

The height of the body is contained twice and three-fourths in the total length
(without caudal); the diameter of the eye is one-fourth of the length of the head, and
less than the extent of the snout, which equals the width of the interorbital space.
The groove for the process of the intermaxillary bones extends beyond the front margin
of the eye, and is entirely naked. The spines of the fins are rather slender; the second
dorsal spine is longer than the third, its length being more than one-half of the height
of the body, and two-thirds of the length of the head; the second of the anal fins is
not so strong as the second of the dorsal, scarcely longer than the third of the anal,
two-sevenths of the height of the body, and two-fifths of the length of the head.
Caudal with scarcely any scales, very deeply forked, the length of its lobes being
contained twice and two-thirds in the total. Colour silvery, with very indistinct
longitudinal interrupted brown stripes on the sides.

Aden. Zanzibar. *East-Indian archipelago.*

Length 8 inches.

372.   **Gerres lineolatus,** sp. n. Plate XVI. fig. 2.   [383 A.]

D. $\frac{9}{10}$. A. $\frac{3}{7}$. L. lat. 38. L. transv. $\frac{6}{11}$.

This species is very closely allied to *G. acinaces.* The height of the body is contained nearly twice and one-half in the total length (without caudal), the length of the
head thrice and one-fourth. The diameter of the eye is contained thrice and a half in
the length of the head, and is slightly less than the snout, which equals the width of
the interorbital space. The snout is more obtuse and less spatulate than in *G. acinaces,*
and the groove for the processes of the intermaxillary bones does not extend quite so
far behind the front margin of the eye.

The spines of the fins are rather slender; the second of the dorsal is longer than
the third, and more than half the height of the body, and five-eighths of the length of
the head. The second of the anal is longer and stronger than the third, and somewhat
stronger than the second, of the dorsal; its length is one-third of the height of the

body, and three-eighths of the length of the head. The caudal lobes are entirely scaleless, and the upper is contained thrice and a quarter in the length of the body.

Colour silvery, with longitudinal lines as in *G. acinaces*, but much more conspicuous. Length 6¾ inches.

Aden. Zanzibar.

373. **Gerres oyena.** [383.]

Labrus oyena, *Forsk.* p. 35.
Smaris oyena, *Rüpp. Atl.* p. 11, t. 3. f. 2.
Gerres oyena, *Cuv. & Val.* vi. p. 472; *Günth. Fish.* iv. p. 261.

Aden. Zanzibar. *Mozambique. Red Sea. Indian Ocean.*

Family CHROMIDÆ.

CHROMIS, *Cuv.*

374. **Chromis niloticus.** [123, 305.]

Labrus niloticus, *Hasselqu.* p. 392.
Chromis niloticus, *Cuv. Règne Anim.; Günth. Fish.* iv. p. 267.

Pangani River. *Nile. Natal.*

Order ANACANTHINI.

Family OPHIDIIDÆ.

FIERASFER, *Cuv.*

375. **Fierasfer neglectus.** [464.]

Fierasfer neglectus, *Peters, Wiegm. Arch.* 1855, p. 260; *Günth. Fish.* iv. p. 382.

The length of the head is one-eighth of the total, and its greatest width is rather more than one-third of the length. Gill-opening very wide, leaving nearly the whole of the isthmus exposed. Vent slightly in advance of the pectorals. Dorsal very inconspicuous; anal low but distinct. Teeth in villiform bands on the jaws, the outer series being somewhat larger; two rather small canines in the front of the upper jaw; three very large, compressed, curved teeth on the vomer.

Colour light, spotted with brown, each spot on the body with a minute black point in the centre; a silvery band along the anterior portion of the lateral line.

Length 3½ inches.

Zanzibar. *Mozambique.*

Family PLEURONECTIDÆ.

PSETTODES, *Benn.*

376.   **Psettodes erumei.**   [433.]

Pleuronectes erumei, *Bl. Schn.* p. 150.
Russell, i. p. 54, pl. 69, and i. p. 60, pl. 77.
Hippoglossus erumei, *Rüpp. Atl. Fische*, p. 121, and *N. W. Fische*, p. 84; *Cant. Mal. Fish.* p. 216.
—— dentex, *Richards. Voy. Sulph. Fish.* p. 102, pl. 47.
? Psettodes belcheri, *Benn. Proc. Comm. Zool. Soc.* 1831, p. 147.
Psettodes erumei, *Günth. Fish.* iv. p. 402.

Zanzibar.   *Red Sea, through all Indian seas, to China.*

PSEUDORHOMBUS, *Bleek.*

377.   **Pseudorhombus russellii.**   [109, 435.]

Platessa russellii, *Gray, Ill. Ind. Zool.*; *Cant. Mal. Fish.* p. 214.
Rhombus oligodon, *Bleek. Act. Soc. Sc. Indo-Nederl.* v. *Japan* v. t. 3. f. 2.
Pseudorhombus russellii, *Günth. Fish.* iv. p. 424.

Aden.   Zanzibar.   *From the east coast of Africa to Australia.*

RHOMBOIDICHTHYS, *Bleek.*

378.   **Rhomboidichthys pantherinus.**   [108, 220, 693.]

Rhombus pantherinus, *Rüpp. Atl. Fische*, p. 121, t. 31. f. 1.
—— parvimanus, *Benn. Proc. Comm. Zool. Soc.* i. p. 168.
Rhomboidichthys pantherinus, *Günth. Fish.* iv. p. 436.

Several specimens, measuring 1¼ inch, have also been collected in the open sea off the east coast of Africa ; we refer them to *R. pantherinus*, because it appears to be the common species of this locality : they are quite colourless, without any trace of scales, but with both eyes already transferred to the left side.   The origin of the dorsal fin is quite as far advanced as in the adult.   Pectorals very small.

Aden.   Zanzibar.   *From the Red Sea and east coast of Africa to the Feejee Islands.*

PARDACHIRUS, *Gthr.*

379.   **Pardachirus marmoratus.**   [125.]

Achirus marmoratus, *Lacép.* iv. pp. 658, 660.
—— barbatus, *Geoffr. Ann. Mus.* i. p. 152, t. 11; *Rüpp. Atl. Fische*, p. 122, t. 31. f. 2.
Pardachirus marmoratus, *Günth. Fish.* iv. p. 478.

Aden.   Zanzibar.   East Africa.   *Madagascar.*

PLAGUSIA, *Cuv.*

380. **Plagusia marmorata.** [448.]

? Plagusia dipterygia, *Rüpp. Atl. Fische*, p. 123, t. 31. f. 3.
Plagusia marmorata, *Bleek. Verh. Batav. Genootsch.* xxiv. *Pleuron.* p. 20, or *Nat. Tydschr. Ned. Ind.* i. p. 411; *Günth. Fish.* iv. p. 491.
Zanzibar. ? *Red Sea. East Indies.*

CYNOGLOSSUS, *Buch. Ham.* *

381. **Cynoglossus quadrilineatus.** [43, 530.]

Achirus bilineatus, *Lacép.* iv. pp. 659, 663.
Plagusia bilineata, *Rüpp. Atl. Fische*, p. 123.
—— quadrilineata, *Bleek. Verhand. Bat. Genootsch.* xxiv. *Pleuron.* p. 21.
Cynoglossus quadrilineatus, *Günth. Fish.* iv. p. 497.
Aden. Zanzibar. *East-Indian archipelago.*

Order PHYSOSTOMI.

Family SILURIDÆ.

CLARIAS, *Gronov.*

382. **Clarias gariepinus.** [187.]

Silurus (Heterobranchus) gariepinus, *Burchell, Travels in the Interior of South Africa,* i. p. 425, c. fig. p. 445.
? Clarias capensis, *Cuv. & Val.* xv. p. 377.
Clarias capensis, *Smith, Illustr. Zool. S. Afr. Fish.* c. tab. (not good).
? Clarias mossambicus, *Peters, Monatsber. Berl. Acad.* 1852, p. 682.
Clarias gariepinus, *Günth. Fish.* v. p. 14.
Fresh water of Zanzibar. *Mozambique. Port Natal. Cape of Good Hope.*

PLOTOSUS, *Lacép.*

383. **Plotosus anguillaris.** [219, 597.]
Aden. Zanzibar. *From East Africa to Polynesia.*

* Undetermined specimen [233].

Q

EUTROPIUS, *Müll. & Trosch.*

384.   **Eutropius,** sp. incerta.   493.]

River Rovuma: immature specimens.

BAGRUS, *Cuv. & Val.*

385.   **Bagrus bayad.**   [499.]

Silurus bajad, *Forsk.* p. 66.

Bayatte, *Sonnini,* pl. 27 (bad).

Porcus bayad, *Geoffr. Desc. Eg. Poiss.* pl. 15. f. 1.

Bagrus bayad, *Cuv. & Val.* xiv. p. 397; *Günth. Fish.* v. p. 69.

Pangani river. East coast of Africa. *Nile. Senegal.*

ARIUS, *Cuv. & Val.*

386.   **Arius thalassinus.**   [120, 151, 407.]

Deddi jella, *Russell,* pl. 169.

Bagrus thalassinus, *Rüpp. N. W. Fische,* p. 75, t. 20. f. 2.

—— bilineatus, *Cuv. & Val.* xiv. p. 434.

—— netuma, *Cuv. & Val.* xiv. p. 438, pl. 417.

Arius nasutus, *Cuv. & Val.* xv. p. 60.

Netuma nasuta, *Bleek. Atl. Ichth.* ii. t. 61.

—— thalassina, *Bleek. Atl. Ichth.* ii. p. 28.

Arius thalassinus, *Günth. Fish.* v. p. 139.

Aden. Zanzibar. *Red Sea. East Indies and East-Indian archipelago.*

387.   **Arius falcarius.**   [501.]

Arius falcarius, *Richards. Voy. Sulph. Fish.* p. 134, pl. 62. figs. 7–9; *Günth. Fish.* v. p. 168.

—— schlegelii, *Bleek. Nederl. Tydschr. Dierk.* 1863, p. 146.

Varietas *africana.*—The African variety of *A. falcarius* differs from the Asiatic one in the following points only. The maxillary barbels extend to the end of the humeral process, the triangular patches of palatine teeth are somewhat larger, the bones of the head a little more coarsely granulated, and the dorsal spine less distinctly serrated behind, which differences cannot justify the creation of a distinct species.

It is not a little remarkable that this species, hitherto only known as existing in the Chinese seas, should be found far up in an East-African river, and never, as far as we are aware, in the salt water of that region; but it is very probable that it will be discovered eventually in some of the intermediate regions.

Pangani river. *Chinese seas.*

SYNODONTIS, *Cuv. & Val.*

388. **Synodontis schal.**

Silurus schall, *Bl. Schn.* p. 385.

Pimelodus clarias, *Geoffr. Descr. Eg. Poiss.* pl. 13. figs. 3, 4.

Synodontis arabi, *Cuv. & Val.* xv. p. 261.

—— schal, *Günth. Fish.* v. p. 212.

—— maculosus, *Rüpp. Beschreib. neuer Nil-Fische*, p. 10, t. 3. f. 1 (young).

Found in the Rovuma river, in the Zanzibar dominions, by Dr. Kirk. *Zambesi. Nile. Senegal.*

389. **Synodontis gambiensis.** Plate XVII. fig. 1. [500.]

Synodontis gambiensis, *Günth. Fish.* v. p. 214.

D. $\frac{1}{7}$. A. 12–13. P. $\frac{1}{8-9}$. V. 7.

The gill-opening extends downwards to before the root of the pectoral fin. Mandibulary teeth much shorter than the eye, in a narrow band. Maxillary barbels much longer than the head, not fringed; the outer mandibulary barbels not much shorter than the head, provided with filaments. The height of the body is contained thrice and a half in the total length (without caudal), and the length of the head four times in the same. Nuchal carapace tectiform, much longer than broad. Dorsal spine rather longer than the head, smooth in front, slightly serrated behind; pectoral spine serrated along both edges, scarcely longer than that of the dorsal fin; humeral process longer than high, pointed behind. The distance between the dorsal and the adipose fin equals the length of the base of the former. Caudal deeply forked, the upper lobe the longer, and contained thrice and three-fifths in the length of the body.

Colour dark brown; head, body, and fins spotted with black.

Length 10½ inches.

Pangani river. East coast of Africa. *Gambia.*

Family SCOPELIDÆ.

SAURUS, *Cuv. & Val.*

390. **Saurus varius.** [231, 562.]

Salmo varius, *Lacép.* v. p. 224, pl. 3. f. 3.

Saurus variegatus, *Quoy & Gaim. Voy. Uran. Poiss.* p. 223, pl. 48. f. 3.

—— varius, *Günth. Fish.* v. p. 395.

Zanzibar. *Indian and Pacific Oceans.*

391. **Saurus atlanticus.** [723.]

Saurus atlanticus, *Johnson, Proc. Zool. Soc.* 1863, p. 41; *Günth. Fish.* v. p. 395.

Br. 15–17. D. 12–13. A. 9–10. L. lat. 59. L. transv. 4/7.

Colour of specimen from Zanzibar in life.—Upper part of body brownish red, variegated with darker and lighter. Two longitudinal series of blue spots more or less

Q 2

confluent into bands above the lateral line, and a continuous similarly coloured band beneath it from base of pectorals to root of caudal. Fins nearly immaculate, some lighter spots on rays of dorsal.

Zanzibar. *Madeira.*

392. **Saurus myops.** [298.]

Salmo fœtens, *Bl.* t. 384. f. 2.
—— myops, (*Forster*) *Bl. Schn.* p. 421.
Saurus myops, *Cuv. Règne Anim.; Cuv. & Val.* xxii. p. 485; *Günth. Fish.* v. p. 398.
—— trachinus, *Schleg. Faun. Japon. Poiss.* p. 231, pl. 106. f. 2; *Cant. Mal. Fish.* p. 271.

Zanzibar. *Tropical parts of Atlantic, Indian, and Pacific Oceans.*

SAURIDA, *Cuv. & Val.*

393. **Saurida nebulosa.** [445.]

Saurus à bandes et taches, *Liénard, Dix. Rapp. Soc. Hist. Nat. Mauric.* 1839, p. 41.
Saurida nebulosa, *Cuv. & Val.* xxii. p. 504, pl. 649; *Günth. Fish.* v. p. 399.

Zanzibar. *Madagascar. Indian Ocean. Western Pacific.*

394. **Saurida tumbil.** [697.]

Badimottah, *Russell,* t. 172.
Salmo tumbil, *Bloch,* ix. p. 112, t. 430.
Saurus badimottah, *Cuv. Règne Anim.; Rüpp. N. W. Fische,* p. 77; *Cant. Mal. Fish.* p. 270.
Saurida tumbil, *Cuv. & Val.* xxii. p. 500; *Günth. Fish.* v. p. 399.

Zanzibar. *Red Sea. Indian Ocean. Western Pacific.*

SCOPELUS, *Cuv.*[*]

395. **Scopelus asper.** [384 b.]

Myctophum asperum, *Richards. Voy. Ereb. & Terr. Ichth.* p. 41, pl. 27. figs. 13–15.
Scopelus asper, *Cuv. & Val.* xxii. p. 454; *Günth. Fish.* v. p. 411.

East coast of Africa. *Gulf of Guinea. New Ireland.*

396. **Scopelus coccoi.** [384 a.]

Scopelus coccoi, *Cocco, in Giorn. Sicil.* fasc. 77, p. 143, and *Lett. s. Salmon.* p. 18, t. 2. f. 6; *Bonap. Faun. Italic. Pesc.* c. fig.; *Günth. Fish.* v. p. 413.
Alysia loricata, *Lowe, Proc. Zool. Soc.* 1839, p. 87, and *Trans. Zool. Soc.* iii. p. 14.
Myctophum hians, *Richards. Voy. Ereb. & Terr. Ichth.* p. 41, pl. 27. figs. 19–21.
? Scopelus jagorii, *Peters, Monatsber. Akad. Wiss. Berl.* 1859, p. 411.

East coast of Africa. *Mediterranean and Atlantic.*

---

* Undetermined specimen [384 c].

Family MORMYRIDÆ.

MORMYRUS, *L.*

397. **Mormyrus macrolepidotus.** [497.]

Mormyrus macrolepidotus, *Peters, Monatsber. Akad. Wiss. Berl.* 1852, p. 275 ; *Günth. Fish.* vi. p. 219.

River Rovuma.

398. **Mormyrus catostoma.**

Mormyrus catostoma, *Günth. Fish.* vi. p. 222.

Found by Mr. C. Livingstone and Dr. Kirk in the river Rovuma.

Family SCOMBRESOCIDÆ.

BELONE, *Cuv.*

399. **Belone choram.** 135.]

Renard, ii. pl. 14. f. 65.

Esox choram, *Forsk. Descr. Anim.* p. 67. no. 98 c.

Belona crocodila, *Lesueur, Journ. Ac. Nat. Sc. Philad.* ii. 1821, p. 129.

Belone choram, *Rüpp. N. W. Fische*, p. 72 ; *Günth. Fish.* vi. p. 239.

—— crocodilus, *Cuv. & Val.* xviii. p. 440.

In the specimens from Zanzibar there are only twenty rays in the dorsal fin.

Aden. Zanzibar. *Mozambique.*

HEMIRAMPHUS, *Cuv.*

400. **Hemiramphus dussumieri.**

Hemiramphus erythrorinchus, var., *Lesueur, Journ. Ac. Nat. Sc. Philad.* ii. p. 138.

? Hemiramphus gamberur, *Rüpp. N. W. Fische*, p. 74.

Hemiramphus dussumieri, *Cuv. & Val.* xix. p. 33 ; *Günth. Fish.* vi. p. 266.

Aden. Zanzibar. Seychelles. *Mozambique. Indian Ocean.*

401. **Hemiramphus commersonii.** [136.]

Hemiramphus commersonii, *Cuv. Règne Anim.* ; *Cuv. & Val.* xix. p. 28 ; *Günth. Fish.* vi. p. 271.

—— far, *Rüpp. N. W. Fische*, p. 74.

Aden. Zanzibar. *Red Sea. Indian Ocean. Mozambique. Natal.*

EXOCŒTUS, *Artedi.*

402. **Exocœtus mento.** [156 n.]

Exocœtus mento, *Cuv. & Val.* xix. p. 124 ; *Günth. Fish.* vi. p. 281.

Zanzibar. *East-Indian archipelago.*

403.   **Exocœtus evolans.**   [150.]
For synonymy see *Günth. Fish.* vi. p. 282.
East coast of Africa.   *Seas of temperate and tropical zones.*

404.   **Exocœtus brachysoma.**   [156 A.]
Exocœtus brachysoma, *Bleek. Nederl. Tydschr. Dierk.* iii. p. 111 ; *Günth. Fish.* vi. p. 295.
Zanzibar.   *Indian and Pacific Oceans.*

Family CYPRINODONTIDÆ.

HAPLOCHILUS, *McCl.*

405.   **Haplochilus playfairii.**   Plate XX. fig. 1.   [332.]
Haplochilus playfairii, *Günth. Fish.* vi. p. 314.

D. 12.   A. 18.   V. 6.   L. lat. 32.   L. transv. 9.

The height of the body is contained four times in the total length (without caudal), the length of the head thrice and a fourth.   Head rather elongate, much depressed anteriorly, the snout being somewhat longer than the eye, which is one-fourth of the length of the head, and more than one-half of the width of the interorbital space.   Jaws equal in length anteriorly.   The origin of the dorsal fin is midway between the extremity of the caudal and the præoperculum, corresponding to the eighteenth scale of the lateral line and to the middle of the anal.   Pectoral fin scarcely extending beyond the root of the ventral, which reaches the anal.   There does not appear to exist a conspicuous difference between the sexes as regards the fins.

Colour in spirits.—Brownish, with three or four indistinct, serrated, dark longitudinal bands.   Sometimes a black line across the base of the middle dorsal rays.

Colour in life.—Yellowish olive, with about seven longitudinal rows of red spots (between the serrated bands mentioned above), corresponding to the series of scales ; opercles with four similarly coloured lines ; vertical fins spotted and reticulated with brown.

Length from 2 to 3½ inches.

Fresh waters of Seychelles.

FUNDULUS, *Lacép.*

406.   **Fundulus orthonotus.**   Plate XVII. figs. 2, 3.   [251, 505.]
Cyprinodon orthonotus, *Peters, Monatsber. Akad. Wiss. Berl.* 1844, p. 35.
Fundulus orthonotus, *Günth. Fish.* vi. p. 326.

D. 15.   A. 15-16.   V. 5.   L. lat. 30-32.   L. transv. 10.

The height of the body is about equal to the length of the head, which is contained thrice and one-third in the total length (without caudal). Head thick, short, with the snout obtuse, the lower jaw projecting beyond the upper. Diameter of the eye about equal to the extent of the snout, one-half of the interorbital space, and two-ninths of the length of the head. The origin of the dorsal fin is midway between the root of the caudal and the eye in males, and equidistant between the root of the caudal and the præoperculum in females. The origin of the anal is opposite to that of the dorsal in males, and somewhat more backwards in females, in which, besides, the anterior anal rays are stiff and inflexible. The male has the vertical fins and the pectorals much more elongate than the females.

Colour in spirits.—Males have all the scales provided with a more or less broad carmine-red margin; sides of the head, dorsal, and anal fin with similarly coloured spots; caudal entirely red. *Females* have the tail and base of the anal and caudal dotted with black.

Colour in life.—Males [251]. Snout yellow, body opalescent, each scale with a carmine-red margin; these are darker and broader behind the origin of the dorsal and on the shoulder, where they assume the appearance of a red patch. Dorsal with bands of reddish brown, much broader than the interspaces between them; upper part blackish; margin white. Anal yellow, with bands similar to those on dorsal. Tail and caudal red, the latter with a black margin. Pectorals yellow, with white margins. Two to four inches in length.

Wells at Zanzibar. Pangani river. Streams at Seychelles. *Quillimane.*

It is remarkable that out of many hundred specimens observed by Colonel Playfair at Zanzibar, Pangani, and Seychelles, no female was ever found at the two first-named places, and no male at the last.

Family CYPRINIDÆ.

LABEO, *Cuv.*

407.　　　　　　　　　　　　**Labeo forskalii.**

Labeo forskalii, *Rüpp. Mus. Senck.* ii. p. 18, tab. 3. fig. 1 ; *Heckel, in Russegger's Reisen,* ii. 3. p. 300, taf. 20. fig. 2.

—— cylindricus, *Peters, Monatsber. Akad. Wiss. Berl.* 1852, p. 684.

River Rovuma. East coast of Africa. *Mozambique. Nile.*

RASBORA, *Bleek.*

408.　　　　　　　**Rasbora zanzibarensis,** sp. n.　Plate XVII. fig. 4.

D. 9.　A. 8.　L. lat. 33.　L. transv. 5½/4.

Dorsal fin inserted behind the ventrals, but in front of the anal. Body compressed, rather elongate, its greatest depth being equal to the length of the head, which is two-

sevenths of the total (without caudal).   Head rather low, with the upper profile somewhat concave.   The width of the interorbital space is much more than the diameter of
the eye, which equals the length of the snout and is one-fourth of that of the head.
Mouth anterior, oblique, of moderate width, the maxillary extending to the vertical
from the front margin of the orbit; lower jaw projecting beyond the upper.   Caudal
fin deeply forked.   Each series of scales along the side of the body with an obtuse
ridge; lateral line much nearer to the ventral than to the dorsal profile.   Coloration
uniform, silvery.   Pharyngeal teeth slender, uncinate, in a triple series.—Length
2¾ inches.

River Rovuma, east coast of Africa.

Family CLUPESOCIDÆ.

CHIROCENTRUS, Cuv.

409.   **Chirocentrus dorab,** *Forsk.*   [372.]

Clupea dorab, *Forsk.* 72. no. 108.
Russell, pl. 199.
Chirocentrus dorab, *Cuv. Règne Anim.*; *Rüpp. N. W. Fische,* p. 81; *Cuv. & Val.* xix. p. 150, pl. 565;
     Cant. Mal. Fish. p. 277.

    Zanzibar.   *Red Sea.   From the east coast of Africa, through Indian Ocean, to Polynesia.*

BUTIRINUS, Comm.

410.   **Butirinus glossodontus.**   [239.]

Argentina glossodonta, *Forsk.* no. 99.
Butyrinus glossodontus, *Rüpp. N. W. Fische,* p. 80, t. 20. f. 3.
Albula bananus, *Lacép.* v. p. 46; *Cuv. & Val.* xix. p. 345.

    Zanzibar.   *Red Sea.*

Family GONORHYNCHIDÆ.

LUTODEIRA, Van Hass.

411.   **Lutodeira chanos.**   [398.]

Mugil chanos, *Forsk.* p. 74. no. 110.
—— salmoneus, *J. B. Foster, Misc.* iv. 14.
Chanos arabicus, *Lacép.* v. p. 395; *Cuv. & Val.* xix. p. 187.
Russell, pl. 207.
Lutodeira chanos, *Rüpp. Atl. Fische,* p. 18, t. 5. f. 1.

Lutodeira salmonea, *Richards. Zool. Voy. Ereb. & Terr. Fish.* p. 58, pl. 36. f. 1.
Chanos nuchalis, *Cuv. & Val.* xix. p. 197.
—— Pala, *Cant. Mal. Fish.* p. 278.
Kiswarra Bay. Seychelles. *All Indian seas.*

Family ELOPIDÆ.

ELOPS, *L.*

412.   **Elops machnata.**   [657, 745.]

Argentina machnata, *Forsk.* p. 68. no. 100.
Russell, pl. 179.
Elops machnata, *Rüpp. N. W. Fische,* p. 80; *Richards. Zool. Voy. Ereb. & Terr. Fish.* p. 59, pl. 36.
    figs. 3–5.
Elops saurus, *Cuv. & Val.* xix. p. 365; *Cantor, Mal. Fish.* p. 287.

The outer rays of the caudal fin, forming its upper and lower margin, are strong and ensiform; young specimens show a remarkable difference from old ones as regards the structure of these rays. The single joints of which they are composed, in young

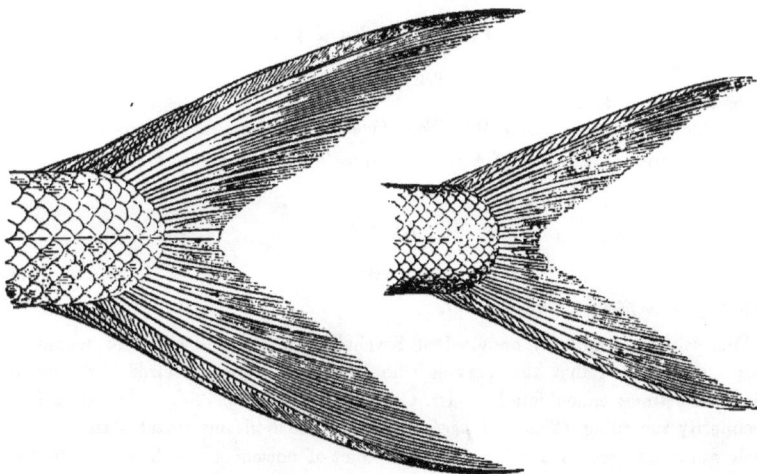

Old; half nat. size.   Young; nat. size.

specimens measuring thirteen inches in length, are of a trapezoid shape, and the total number of those rays is from thirty-two to thirty-five. In old specimens of upwards of

R

three feet in length, the joints are linear, very obliquely placed, and their number is from seventy to ninety. We thought it well to make this observation, as it may throw some light on the growth and regeneration of fin-rays.

Zanzibar. *Red Sea. Indian Ocean and Archipelago.*

413.   **Elops cyprinoides.**   [169.]
Clupea cyprinoides, *L. Gm.*
—— apalike, *Bonnat. Encycl. Méth.*; *Lacép.* v. p. 459, t. 13. f. 3.
Russell, pl. 203.
Cyprinodon cundinga, *Buch. Ham.* 254, 383.
Megalops filamentosus, *Cuv. Règne Anim.*
—— indicus, *Cuv. & Val.* xix. pp. 388, 577.
Elops indicus, *Cuv. & Val.* xx. p. 472.
—— cundinga, *Cant. Mal. Fish.* p. 289.

Pangani river. East coast of Africa. *Both coasts of Africa. India. Ceylon. China. Polynesia.*

Family CLUPEIDÆ.

PELLONA, *Cuv. & Val.*

414.   **Pellona ditchoa.**   [655.]
Ditchoa, *Russell*, pl. 192.
Pellona ditchoa, *Cuv. & Val.* xx. p. 313; *Bleek. Verhand. Bat. Genootsch.* xxiv. *Har.* p. 24.

Kingani river. East coast of Africa. *Indian Ocean and Archipelago.*

ALOSA.

415.   **Alosa venenosa.**   [103.]
Meletta venenosa, *Cuv. & Val.* xx. p. 377.

This fish was originally obtained at Seychelles by M. Dussumier, in whose notes there is a remark that the persons who partook of it were seized with vomiting, which sometimes ended fatally. Dr. Cantor (Mal. Fish. p. 295) mentions a similar peculiarity regarding *Clupeonia perforata*, Cant.; and his informant stated that only such fish as had red eyes induced the symptoms of poisoning; such as had the usual silvery eyes were eaten with impunity. The specimens of this collection were obtained at Zanzibar, where they do not appear to possess any such property; on the contrary, they form no inconsiderable part of the daily food of the lower orders.

Zanzibar. *Seychelles. Indian Ocean.*

416. **Alosa chapra.** [104, 656.]

Alosa chapra, *Gray, Ill. Ind. Zool.* pl. 92. f. 2; *Cuv. & Val.* xx. p. 440.

D. 17. A. 21. L. lat. 44.

The height of the body is one-third of the length (to the fork of caudal); the length of the head is contained thrice and two-thirds in the same. Snout rather longer than the diameter of the eye. Operculum twice as high as long. Belly strongly compressed, very sharp, but not very distinctly serrated. The origin of the dorsal is nearer the end of the snout than the root of the caudal; its last ray is very slightly produced, longer than the two or three before it, but not half that of the longest one, which is contained twice and two-thirds in the height of the body. The anal is very low, and occupies the distance of ten transverse series of scales. Root of the pectorals below the middle of the dorsal. Colour blue above, silvery below. In adults there is a very distinct round blackish spot on the shoulder, behind the operculum, which is inconspicuous in young specimens.

Aden. Zanzibar. *East Indies.*

417. **Alosa kowal.** [746.]

Clupea kowal, *Russell,* pl. 186; *Rüpp. N. W. Fische,* p. 79.; *Schleg. Faun. Japon.* p. 235, pl. 107. f. 1. Kowala thoricata, *Cuv. & Val.* xv. p. 363; *Cant. Mal. Fish.* p. 296.

Zanzibar. Aden. *Red Sea. Indian Ocean.*

418. **Alosa punctata.** [745 a.]

Clupea punctata, *Rüpp. N. W. Fische,* p. 78, t. 21. f. 2.

Aden. Zanzibar. *Red Sea.*

419. **Alosa sirm.** [414.]

Clupea sirm, *Forsk.* p. xvii. no. 44; *Rüpp. N. W. Fische,* p. 77, t. 21. f. 1.

Zanzibar. *Mozambique. Red Sea.*

ENGRAULIS, *Cuv.*

420. **Engraulis brownii.** [656.]

Atherina brownii, *L. Gm.* 1397.
Russell, pl. 187.
Engraulis fasciata, *Cuv. & Val.* xxii. p. 43.
—— brownii, *Cuv. & Val.* xxii. p. 41; *Cant. Mal. Fish.* p. 303.

Zanzibar. *Nearly all tropical seas.*

421. **Engraulis boelama.** [631.]

Clupea boelama, *Forsk.* p. 72. no. 107.
Engraulis boelama, *Cuv. & Val.* xxi. p. 55.

Zanzibar. *Seychelles. Réunion. Red Sea.*

Order APODES.

Family ANGUILLIDÆ.

ANGUILLA, *Thunb.*

422.   **Anguilla johannæ,** sp. n.   [574.]

Similar in habit to *A. maculata*, Buch. Ham., and to *A. labiata*, Pet., but differing from them in the relative length of the distance between the origin of dorsal and anal.

Body rather stout. Eye moderate, placed entirely in front of the cleft of the mouth; its diameter is less than half the breadth of the interorbital space, and is contained about twice and a half in the length of the snout. Snout of moderate length and width, depressed. The lower jaw distinctly longer than the upper. Lateral lips moderately developed. Lateral teeth in two series; the vomerine patch acutely tapering behind, but not extending so far back as the lateral series do. Pectorals rather longer than the cleft of the mouth. The distance between the point of the snout and the base of the pectorals is about equal to the distance thence to the origin of the dorsal, or is contained about seven and a half times in the total length. The distance between the vent and the vertical from the origin of the dorsal is contained *twice and a third* in the length of the body in front of the vent (in *A. maculata* it is *twice and two-thirds*, and in *A. labiata* it is *thrice and a half*). Dorsal and anal moderately developed. Colour brown, marbled with darker and lighter.

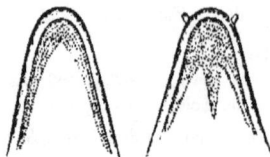

Length 28 inches. Fresh water of Island of Johanna.

423.   **Anguilla labiata.**   [503.]

Anguilla labiata, *Peters, Wiegm. Archiv,* 1855, p. 270; *Kaup, Cat. Ap. Fish.* p. 41.

General habit of body rather stout; eye moderate, placed entirely in advance of the angle of the mouth; its diameter is about half the width of the interorbital space, or of the length of the snout. Snout short, broad, obtuse, depressed, the lower jaw slightly longer than the upper. Lateral lips much developed, broad. Teeth on the sides of the jaws in three or four series; the vomerine patch is tapering behind, and does not reach as far back as the lateral teeth do. Pectorals as long as the cleft of the mouth. The distance from the end of the snout to the base of the pectorals is equal to that from the pectorals to the origin of the dorsal, or about one-eighth of the total length. The vent is placed somewhat behind the origin of the dorsal. Colour greenish brown, lighter on the belly; pectorals blackish.

Length 3 feet 6 inches.

Pangani River. East coast of Africa. *Mozambique.*

424.   Anguilla amblodon, sp. n.   [658.]

General habit moderately slender. Eye small, much less than half the width of the interorbital space, and about one-fifth of the length of the snout. Snout obtuse; the lower jaw somewhat prominent. Teeth in many series in both jaws; the vomerine patch is rounded behind, and extends as far back as the centre of the eye. Lips moderately thick. Pectorals not much shorter than the mandible. The length between the base of the pectorals and the end of the snout is two-fifths of the distance from the anterior of the snout to the base of the pectoral; it about equals the distance thence to the origin of the dorsal, or about one-eighth of the total length of the body. The distance between the vent and the vertical from the origin of the dorsal is about one-third of the length of the body in front of the vent. A very distinct series of pores along the lateral line. Dorsal and anal well developed; the highest portion of the former is about half the height of the body beneath it. Colour above greenish, marbled with dark brown; the lower portion of the body is of a dirty white colour.

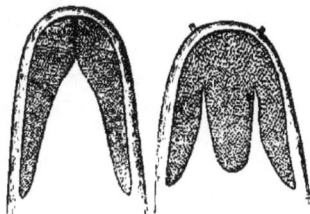

Fresh water of Seychelles.

Length about 2 feet.

Family CONGRIDÆ.

CONGER, Cuv.

425.   **Conger altipinnis.**   [606, 685.]

? Muræna tota cinerea, *Forsk.* p. 22. no. 9.
? Conger cinereus, *Rüpp. Atl. Fische,* p. 115, t. 29. f. 1.
Conger altipinnis, *Kaup, Cat. Ap. Fish.* p. 114.
——— noordzicki, *Bleek. Atl. Ichth.* iv. p. 26, t. 167. f. 2.

Amongst the numerous examples of this fish from Zanzibar, there is a banded specimen which agrees tolerably well with Rüppell's figure of *C. cinereus,* except that in the latter the dorsal fin originates behind the termination of the pectorals, whereas in all the Zanzibar specimens, and according to Kaup's and Bleeker's descriptions, the pectoral reaches beyond the commencement of the dorsal.

Zanzibar. *Mozambique. Réunion. East-Indian archipelago. Red Sea.*

Family MURÆNIDÆ.

MURÆNA, *Thunberg**.

426. **Muræna chlorostigma.** [086.]

Thyrsoidea chlorostigma, *Kaup, Cat. Ap. Fish.* p. 89.
Gymnothorax chlorostigma, *Bleek. Atl. Ichth.* iv. p. 97, t. 178. f. 2.

Zanzibar. Seychelles. *East-Indian archipelago.*

427. **Muræna tessellata.** [592.]

Muræna tessellata, *Rich. Zool. Voy. Sulph. Fish.* p. 109, t. 55. f. 5–8, and *Zool. Voy. Ereb. & Terr. Fish.* p. 88.
Gymnothorax tessellatus, *Bleek. Atl. Ichth.* p. 93, pl. 171. f. 3, and pl. 172. f. 1.

The spots in this specimen are almost perfectly hexagonal, and separated by narrow meshes of bright yellow.

Zanzibar. *East-Indian archipelago.*

428. **Muræna isingleena.** [437.]

Muræna isingteena *and* isingleena, *Richards. Ichth. China, Rep.* 15*th Meet. Brit. Assoc.* p. 314, and *Zool. Voy. Sulph. Fish.* p. 108, pl. 48. f. 1, and *Ichth. Voy. Ereb. & Terr.* p. 86.
Thyrsoidea isingleena, *Kaup, Cat. Ap. Fish.* p. 75, pl. 11. f. 57.
Gymnothorax isingteena, *Bleek. Atl. Ichth.* iv. p. 92, t. 181. f. 1.

In the specimens from Zanzibar the spots had no white edges.

Zanzibar. *East-Indian archipelago. China Seas.*

429. **Muræna isingleenoides.** [687.]

Muræna isingleena, *Bleek. Nat. Tydsch. Ned. Ind.* ix. *Sumatra,* p. 227 (not *Richards.*).
Gymnothorax isingleenoides, *Bleek. Atl. Ichth.* iv. p. 91, t. 179. f. 1, and t. 180. f. 1.

Zanzibar. *East-Indian archipelago.*

430. **Muræna picta.** [41, 462, 605, 717.]

Muræna picta, *J. N. Ahl de Muræn. et Ophichth. in Thunb. Diss.* iii. p. 6, t. 2. f. 2.
Murænophis pantherina, *Lacép.* v. pp. 628, 641, 643.
Muræna variegata, *Quoy & Gaim. Voy. Uran. Poiss.* t. 52. f. 1; *Richards. Zool. Voy. Ereb. & Terr. Fish.* p. 84.
Muræna siderea, *Richards. Zool. Voy. Ereb. & Terr. Fish.* pp. 84,85, t. 48. f. 1–5.
Gymnothorax pictus, *Bleek. Atl. Ichth.* iv. p. 87, t. 170. f. 3, 4, t. 172. f. 3, t. 173. f. 1, and t. 189. f. 3.

Of all the specimens from Zanzibar, only one [717] has the larger dark round spots.

Zanzibar. Aden. *East-Indian archipelago.*

* Undetermined specimen [451].

431. **Muræna nubila.** [107, 461, 463, 535.]

Muræna nubila, *Richards. Zool. Ereb. & Terr. Fish.* p. 81, pl. 46. f. 6–10; *Kaup, Cat. Ap. Fish.* p. 57, fig. 48.

Aden. Zanzibar. *Mauritius. Norfolk Island.*

432. **Muræna flavomarginata.** [718.]

? Murenophis grise, var. *Lacép.* v. 641, pl. 19. f. 3.
Muræna flavimarginata, *Rüpp. Atl. Fische,* p. 119, t. 30. f. 3.
Zanzibar. Mozambique. *Red Sea.*

433. **Muræna nudivomer,** sp. n. Plate XVIII. [460, 551.]

Vomerine teeth none; upper jaw with a cluster of five or six slightly curved teeth in front (the hindermost being the longest), and with from sixteen to eighteen rather strong teeth in the lateral series. These numbers, however, are not constant, as teeth are constantly being shed. The distance of the gill-opening from the extremity of the snout is one-eighth or one-ninth of the total length, and twice the length of the mandible. Origin of dorsal fin opposite the middle of the distance between gill-opening and mandibulary joint. Dorsal and anal fins very distinct and of moderate height. General habit of body rather slender.

Colour in a dried state.—Ground-colour pale reddish yellow; head and anterior parts of trunk densely covered with small irregular brown specks, which, towards the hind part of the trunk, become confluent into vermiculated lines. On the tail the lines form a regular network enclosing oval spots, each about twice the size of the eye. The fins participate in the coloration of the corresponding parts of the body. Length of typical specimen 2 feet 10 inches.

In another specimen, hardly shorter than the other, the vermiculated lines commence before the middle of the trunk, forming a network on the hind portion; on the tail the brown becomes the ground-colour; this part is beautifully ornamented with ovate, yellow, black-edged ocelli, each the size of the eye.

Zanzibar.

This species resembles *Muræna ocellata*, Agassiz (Pisc. Bras. p. 91, t. L. b. figs. 6–9. and Richards. Voy. Ereb. & Terr. Zool. Fish. p. 82, pl. 47. figs. 6–10), but is distinguished from it by the absence of vomerine teeth, and by a marked difference in the coloration; the Mexican species has light markings on a dark ground, on the fore part of the fish, while that from Zanzibar has dark markings on a light ground.

Family OPHIURIDÆ.

OPHIURUS, Lacép.

434.   **Ophiurus marginatus.**   [378.]

Ophiurus marginatus, *Peters, Wiegm. Arch.* 1855, p. 272.

Body slender and cylindrical; the length of the præanal part is contained once and a third in the tail. Tail scarcely compressed, tapering to a point. The length of the head to gill-openings is contained nearly seventeen times in the total length, and seven times in the length to the vent. The distance between the angle of the mouth and the end of the snout is one-third of the length of the head. Eye small, its diameter being one-third of the length of the snout in front of it. The upper jaw projects considerably beyond the lower one. The dorsal commences slightly behind the extremity of the pectorals; it is rather low and is received into a deep groove, as is also the anal. Pectorals short, their length being contained five and a half times in that of the head in front of the gill-openings.

Colour yellowish, the upper portion of the fish minutely and densely punctulated with black.

Length 1 foot 1½ inch.

Open sea, east coast of Africa. *Mozambique.*

Family LEPTOCEPHALIDÆ.

LEPTOCEPHALUS, L.

Although we attach names to the specimens collected on the east coast of Africa, we do not mean to express ourselves in favour of the views of those who regard the *Leptocephalidæ* as perfect animals. On the contrary, we are inclined to regard them as larval forms of some fish. In the present state of our knowledge it is hardly safe to refer them to the eels; it is much more probable that they are larvæ of some deep-sea fish as yet unknown.

We think we have recognized the following forms amongst the specimens of this collection.

435.   **Leptocephalus marginatus.**   [708.]

Leptocephalus marginatus, *Kaup, Cat. Ap. Fish.* p. 152, pl. 19. f. 19.

East coast of Africa. *Coast of India.*

436.   **Leptocephalus dentex.**   [692.]

Leptocephalus dentex, *Cant. Mal. Fish.* p. 333 ; *Kaup, Cat. Ap. Fish.* p. 151.
  East coast of Africa.   *Penang.*

437.   **Leptocephalus capensis.**   [677.]

Leptocephalus capensis, (*Lalande*) *Kaup, Cat. Ap. Fish.* p. 153.
  East coast of Africa.   *Cape of Good Hope.*

Order PLECTOGNATHI.

OSTRACION, *L.*

438.   **Ostracion turritus.**   [82.]

Ostracion turritus, *Forsk.* p. 75. no. 113 ; *Bl.* i. p. 113, pl. 136 ; *Rüpp. N. W. Fische,* p. 5.
—— (Tetrosomus) turritus, *Bleek. Atl. Ichth.* v. p. 31, pl. 203. f. 3.
  Aden.   Zanzibar.   *Red Sea.   East-Indian archipelago.*

439.   **Ostracion arcus.**   [39.]

Seb. iii. pl. 24. f. 8, 9, 13.
Ostracion cornutus, *Bl.* i. p. 105, t. 133 ; *Schleg. Faun. Japon. Poiss.* p. 229, t. 131. f. 4 (young).
—— arcus, *Bl. Schn.* p. 502.
Ostracium cornutum, *Cant. Mal. Fish.* p. 365.
Ostracion (Acanthostracion) arcus, *Bleek. Atl. Ichth.* v. p. 35, pl. 202. f. 3, and 204. f. 4.
  Aden.   Zanzibar.   Seychelles.   *East-Indian archipelago.   Japan.   Indian Seas.*

440.   **Ostracion fornasini.**   [314.]

Ostracion fornasini, *Bianc. Spec. Zool. Mos.* fasc. i. p. 7, and *Mem. Accad. Sc. Bologn.* vi. p. 151,
  pl. 1. f. 1.
—— pentacanthus, *Bleek. Act. Soc. Scient. Indo-Nederl.* ii. *Amboina,* p. 98.
—— (Acanthostracion) fornasini, *Bleek. Atl. Ichth.* v. p. 34, t. 203. f. 4.

Dr. Peters (Wiegm. Archiv, 1855, p. 276) states that he considers this to be only an
abnormal specimen of *O. cornutus,* Bl. ; but the figures both of Bianconi and of Bleeker
exactly correspond with the specimens obtained at Zanzibar.
  Zanzibar.   *Mozambique.   East-Indian archipelago.*

441.   **Ostracion tetragonus.**   [121.

Ostracion, sp., *Artedi, Syn.* p. 84. no. 6, and p. 85. no. 8 ; and *Gen.* p. 39. no. 4.

s

130 PLECTOGNATHI.

Ostracion tetragonus, *L. Mus. Ad. Frid.* p. 59; *Bleek. Atl. Ichth.* v. p.39, pl. 201. f. 2, and pl. 203. f. 2.
—— tuberculatus, *L. Syst. Nat.* ed. 10. i. p. 331.
—— gibbosus, *L. l. c.* p. 332.
—— cubicus, *L. l. c.*; *Bl.* i. p. 115, t. 137; *Rüpp. Atl. Fische*, p. 3; *Lefebv. Voy. Poiss.* pl. 8; *Peters, Wiegm. Arch.* 1855, p. 275.
Abu senduk, *Forsk.* p. 17. no. 48.
Ostracion cyanurus, *Rüpp. l. c.* p. 4, t. 1. f. 2.
—— immaculatus, *Schleg. Faun. Japon.* p. 296.
Aden. Zanzibar. Mozambique. *From Red Sea, through Indian Ocean, to Japan.*

442. **Ostracion punctatus.** [382.]
Seba, iii. p. 61, t. 24. f. 5.
Ostracion punctatus, *Lacép.* i. pp. 442, 455, t. 21. f. 1; *Bleek. Atl. Ichth.* v. p. 39, pl. 202. f. 4.
Zanzibar. *Mauritius. East-Indian archipelago. Tahiti. New Holland.*

443. **Ostracion bombifrons.** [287.]
Ostracion bombifrons, *Hollard, Ann. Sc. Nat.* 4ᵉ ser. vii. p. 168.
—— sebæ, *Bleek. Atl. Ichth.* v. p. 41, pl. 214. f. 1 (not *Seba*, iii. t. 24. f. 5).
Zanzibar. *East-Indian archipelago.*

DIODON, L.

444. **Diodon reticulatus.** [318.]
Orbis muricatus *et* reticulatus. *Will.* tab. J. no. 7, p. 155; *Seba*, iii. p. 58, t. 23. f. 3.
Ostracion subrotundus, *Artedi, Gen.* p. 52. no. 16; *Synon.* p. 86. no. 19.
Diodon reticulatus, *L. Syst. Nat.* ed. 10. i. p. 334.
Chilomycterus reticulatus, *Bibr. ap. Brisout de Barneville, Note sur les Diod. Revue Zool.* 1846, p. 142; *Bleek. Atl. Ichth.* v. p. 54.
Dicotylichthys punctulatus, *Kaup, Arch. Natury.* i. p. 230.
Zanzibar. *East-Indian archipelago.*

TETRODON, L.

445. **Tetrodon honkenii.** [596.]
Tetrodon honkenji, *Rüpp. Atl. Fische*, p. 65, t. 17. f. 2.
—— hypselogenion, *Bleek. Atl. Ichth.* v. p. 61, pl. 213. f. 5.
Zanzibar. *Red Sea. East-Indian archipelago.*

446. **Tetrodon argenteus.** [68. 315.]
Tetraodon argenteus, *Lacép. Ann. Mus. d'Hist. Nat.* iv. p. 211, t. 58. f. 2; *Schleg. Faun. Japon. Poiss.* p. 275, t. 121. f. 2; *Bleek. Atl. Ichth.* v. p. 64, pl. 209. f. 1.
—— argyropleura, *Benn. Proc. Comm. Zool. Soc.* 1832, p. 184.

[315] is an immature specimen, in which the spots on the back are small, crowded, and somewhat indistinct.

Cape Guardafui. Zanzibar. *Seychelles. Red Sea. India. East-Indian archipelago. Japan. New Holland.*

447.   **Tetrodon lunaris.**   [744.]

Russell, i. p. 20. f. 29.

Tetrodon lunaris, *Bl. Schn.* p. 505 ; *Cuv. Règne An.* ; *Cant. Mal. Fish.* p. 378 ; *Bleek. Atl. Ichth.* v. p. 63, pl. 205. f. 2.

—— lepa, *Buch. Ham.* pp. 10, 362.

—— leropleura, *Gray, Illust. Ind. Zool.* i. *Pisces,* t. 4. f. 2.

Zanzibar. *From Red Sea to Japan.*

448.   **Tetrodon lineatus.**

Synonymy of the adult fish [317] :—

Tetraodon mappa, *Less. Zool. Voy. Coqu. Poiss.* pl. 5.

Synonymy of young examples [628, 715] :—

Tetrodon lineatus, *Bl.* i. p. 128, t. 141 ; *Peters, Wiegm. Archiv,* 1855, p. 274 (not *L.* or *Lacép.*).

—— aerostatus, *Jen. Zool. Voy. Beagle, Fish.* p. 152 ; *Schleg. Faun. Japon. Poiss.* 287, t. 125. f. 2 & 3. Crayracion lineatus, *Bleek. Atl. Ichth.* v. p. 70, pl. 206. f. 1.

Zanzibar. *Mozambique. East-Indian archipelago. New Guinea. Japan.*

449.   **Tetrodon laterna.**   [38.]

?? Tetrodon implutus, *Jen. Zool. Voy. Beagle, Fish.* p. 152.

? —— laterna, *Richards. Voy. Sulph. Zool. Fish.* p. 124, pl. 61. f. 2.

? Crayracion laterna, *Bleek. Atl. Ichth.* v. pl. 205. f. 3.

? —— implutus, *Bleek. l. c.* p. 71.

Our specimens agree with Bleeker's figure as regards general habit, and consequently may be identical with his fish.

The width of the bony part of the interorbital space is not much less than the extent of the snout, and is conspicuously concave. The specimens (from Aden and Zanzibar) agree perfectly as regards the spiny nature of the skin ; the spines extend superiorly from the interorbital space to the middle of the distance between the dorsal and caudal. and inferiorly from the chin to the anal. They are rather prominent and densely crowded on the belly, and less prominent on the back and sides.

The chief difference between our specimens and those described by other authors is the coloration. In the former the upper parts are brown, with white ocelli ; belly uniform yellowish white ; sides with or without several longitudinal white interrupted bands, but always with four irregular black blotches—one at some distance below the eye, the

second below and somewhat in front of the pectoral, the third below and behind that
fin, and the last at some distance behind the third. The curved white lines round the
orbit and base of pectorals are as described by Bleeker.

Aden. Zanzibar. *Red Sea. East-Indian archipelago.*

450.   **Tetrodon stellatus.**   [725.]

Tétrodon étoilé, *Lacép.* i. pp. 474, 483; *Eyd. & Soul. Zool. Voy. Bonite, Atl. Poiss.* t. 10. f. 2 (var.).
—— commersonii, *Bl. Schn.* p. 508.
—— punctatus, *Bl. Schn.* p. 506.
—— lagocephalus, *Bl. Schn.* p. 503.
Calamara kappa, *Russell*, pl. 28 (var.).
Tetrodon calamara, *Rüpp. Atl. Fische*, p. 64, t. 17. f. 1 (var.), and *N. W. Fische*, p. 61.
Tétrodon panthère, *Eyd. & Soul. l. c.* t. 10. f. 3.
Crayracion stellatus, *Bleek. Atl. Ichth.* v. p. 73, t. 209. f. 2.

Zanzibar. *Mauritius. Red Sea. Coromandel. Sandwich Islands. New Ireland.*

451.   **Tetrodon nigropunctatus.**   [276.]

Tetrodon nigropunctatus, *Bl. Schn.* p. 507.
Crayracion nigropunctatus, *Bleek. Atl. Ichth.* v. p. 74, tab. 206. f. 4.

Zanzibar. *East-Indian archipelago.*

452.   **Tetrodon immaculatus.**   [37, 469.]

Tetrodon immaculatus, *Lacép.* i. p. 475, 486, t. 24. f. 1; *Cant. Mal. Fish.* p. 373.
Kappa, *Russell*, i. p. 19. f. 26.
Tetrodon sordidus, *Rüpp. N. W. Fische*, p. 60, t. 16. f. 4.
—— scaber, *Eyd. & Soul. Voy. Bonite, Zool.* p. 214; *Atl. Poiss.* t. 10. f. 1.
Crayracion immaculatus, *Bleek. Atl. Ichth.* v. p. 75, pl. 211. f. 1.

Aden. Zanzibar. *Seychelles. Réunion. Mauritius. Red Sea. East-Indian archipelago.*

453.   **Tetrodon valentyni.**   [399.]

Ikan kakassee, *Valent. Ind. orient. Amb.* p. 353. f. 19, and p. 408. f. 195.
Tropidichthys valentynii, *Bleek. Nat. Tydschr. Nederl. Ind.* iv. *Amb.* iii. p. 130.
Holocanthus balistæformis, *Gron. Cat. Fish. ed. Gray*, p. 25.
Tetrodon tæniatus, *Peters, Wiegm. Archiv*, 1855, p. 275.
Canthogaster valentyni, *Bleek. Atl. Ichth.* v. p. 80, pl. 208. f. 1.

The specimens from Zanzibar differ somewhat in coloration from those previously
described. The general colour of the upper part, between the bands, is yellowish;
the belly white; the snout is brownish, with numerous darker longitudinal lines
above; four short blue lines issue from the posterior margin of the eye; the sides of
the head have numerous round brown spots.

The body has *four* dark-brown cross bands:—the first behind the eye; the second over the middle of the back, interrupted at the pectoral, and continued below it towards the belly, the two portions forming nearly a right angle; the third from the base of the dorsal and before it towards the belly; the fourth on the upper part of the tail. These bands are marked with indistinct bluish streaks, and have deep-brown spots; the interstices between them have numerous brown ocelli; none, however, occur on the belly. Caudal yellow, with dark upper and lower margins, and a few brown ocelli on the base.

Zanzibar. *Mozambique. East-Indian archipelago.*

454. **Tetrodon margaritatus.** [81.]

Tetrodon margaritatus, *Rüpp. Atl. Fische*, p. 66.
—— cinctus, *Sol. ap. Richards. Zool. Voy. Sulph. Fish.* p. 125.
—— solandri, *Richards. l. c.* p. 125, t. 54. f. 4-6.
—— insignatus, *Richards. Zool. Voy. Samar. Fish.* p. 20, pl. 9. f. 1, 2.
—— ocellatus, *Peters, Wiegm. Archiv*, 1855, p. 274.
—— petersii, *Bianc. Spec. Zool. Mosamb.* p. 225, t. 2. f. 1, 2.
Canthogaster margaritatus, *Bleek. Atl. Ichth.* v. p. 81, t. 213. f. 4.

Aden. Zanzibar. *Mozambique. East-Indian archipelago.*

ERYTHRODON, *Bleek.*

455. **Erythrodon niger.** [286.]

Balistes niger, *Lacép.* i. p. 378, pl. 15. f. 2.
Xenodon niger, *Rüpp. N. W. Fische*, p. 53, t. 14. f. 3.
Erythrodon niger, *Bleek. Verhand. Bat. Genootsch.* xxiv. *Balist.* p. 37, and *Atl. Ichth.* v. pl. 219, and *Ned. Tydschr. Dierk.* iii. 1865, p. 21.

Mombassa. *Red Sea. East-Indian archipelago. Réunion.*

BALISTES, *L.*

456. **Balistes cærulescens.** [492.]

Balistes cærulescens, *Rüpp. Atl. Fische*, p. 32, t. 7. f. 2, and *N. W. Fische*, p. 60.

Zanzibar. *Red Sea.*

457. **Balistes frenatus.** [88.]

Balistes frenatus, *Lacép.* i. p. 378, pl. 14. f. 3; *Bleek. Atl. Ichth.* v. pl. 223. f. 2, and *Ned. Tydschr. Dierk.* iii. 1865, p. 22.
—— amboinensis, *Gray, Ill. Ind. Zool.* i. pl. 90. f. 2.

Balistes hihpe, *Richards. Ichth. Voy. Sulph.* p. 127, pl. 59. f. 2.
—— schmittii, *Bleek. Verhand. Bat. Genootsch.* xxiv. *Balist.* p. 37, and *Nat. Tydschr. Ned. Ind. v.* p. 532.

Aden. Zanzibar. *Madagascar. Mozambique. New Caledonia. East-Indian archipelago. China. Polynesia.*

458. **Balistes armatus.** [199, 346.]

Balistes armatus, *Lacép.* i. p. 378, t. 18. f. 2.
—— chrysopterus, *Bl. Schn.* p. 466.
—— niger, *Mungo Park, Trans. Linn. Soc. Lond.* iii. p. 33.
—— subarmatus, *Gray, Ill. Ind. Zool.*
—— albicaudatus, *Rüpp. N. W. Fische,* p. 54, t. 16. f. 1.
—— (Balistapus) armatus, *Bleek. Atl. Ichth.* v. pl. 216. f. 1, and *Ned. Tydschr. Dierk.* iii. p. 23.

Zanzibar. Johanna. *Red Sea. Mozambique. Indian Ocean and Archipelago. China.*

459. **Balistes aculeatus.** [319.]

Klein, Miss. iii. t. 13. f. 10; *Seba,* iii. t. 24. f. 15.
Balistes aculeatus, *L. Syst. Nat.* p. 406; *Bl.* t. 149; *Lay & Benn. Beechey's Voy.* p. 69, pl. 22. f. 2.
—— spinosus, *Lacép.* i. p. 367, pl. 17. f. 1.
—— ornatissimus, *Less. Voy. Coqu. Poiss.* pl. 10. f. 1.
—— armatus, *Cuv. Règne Anim. Ill.* pl. 112. f. 2.
Balistes (Balistapus) aculeatus, *Bleek. Verh. Bat. Genootsch.* xxiv. *Balist.* p. 15; *Bleek. Atl. Ichth.* v. pl. 216. f. 3, and *Ned. Tydschr. Dierk.* iii. p. 24.

Zanzibar. *Red Sea. Mozambique. Mauritius. Indian Ocean and Archipelago. Australia.*

460. **Balistes lineatus.** [345.]

Balistes aculeatus minor, *Forsk.* no. 47.
—— lineatus, *Bl. Schn.* t. 87; *Bleek. Verhand. Bat. Genootsch.* xxiv. *Balist.* p. 14.
—— undulatus, *Mungo Park, Trans. Linn. Soc. Lond.* iii. p. 33.
—— aculeatus *b.* viridis, *Benn. Fishes of Ceylon,* pl. 10.
—— lamourouxii, *Quoy & Gaim. Voy. Uran.* pl. 47. f. 1 (not good).
—— sesquilineatus, *Lay & Benn. Beechey's Voy.* p. 69, pl. 21. f. 3.
Balistes (Balistapus) lineatus, *Bleek. Atl. Ichth.* v. pl. 229. f. 2, and *Ned. Tydschr. Dierk.* iii. 1865, p. 24.

Johanna. *Red Sea. Mozambique. East-Indian Ocean and Archipelago. China. Japan. Tahiti.*

461. **Balistes rectangulus.** [720.]

Balistes écharpe, *Lacép.* i. p. 352, t. 16. f. 1.
—— rectangulus, *Bl. Schn.* p. 465.

Balistes medinilla, *Quoy & Gaim. Voy. Uran. Poiss.* pl. 46. f. 2 (not good).
—— erythropteron, *Less. Voy. Coqu. Poiss.* pl. 10. f. 3.
—— cinctus, *Bleek. Act. Soc. Scient. Indo-Nederl.* ii. *Amb.* p. 96.
Balistes (Balistapus) cinctus, *Bleek. Atl. Ichth.* v. pl. 228. f. 1, and *Ned. Tydschr. Dierk.* iii. 1865, p. 24.

Mozambique. *Red Sea. Indian Ocean and Archipelago. China. Japan. Polynesia.*

462. **Balistes stellatus.** [200.]

Balistes stellatus, *Lacép.* i. p. 350, t. 15. f. 1.
—— stellaris, *Bl. Schn.* p. 476.
—— brasiliensis, *Bl. Schn.* p. 470.
? *Russell,* pp. 22, 23.
Balistes vachellii, *Richards. Zool. Voy. Sulph. Fish.* p. 129.
—— phalliatus, *Richards. Ichth. China.*
Leiurus stellatus, *Bleek. Atl. Ichth.* v. pl. 215, and *Ned. Tydschr. Dierk.* iii. 1865, p. 21.

Zanzibar. *Mauritius. Mozambique. Red Sea. Indian Ocean and Archipelago. New Ireland.*

463. **Balistes viridescens.** [218.]

Balistes aculeatus major, *Forsk.* no. 46.
—— viridescens, *Lacép.* i. p. 378; *Bl. Schn.* p. 477.
—— castaneus, *Richards. Ichth. Voy. Sulph.* p. 126, pl. 59. figs. 5 & 6.
Balistes (Pseudobalistes) viridescens, *Bleek. Atl. Ichth.* v. t. 231. f. 2; and *Ned. Tydschr. Dierk.* iii. 1865, p. 21.

Zanzibar. *Mauritius. Red Sea. East-Indian archipelago.*

464. **Balistes niger.** Plate XIX. fig. 1. [163.]

Balistes niger, *Osbeck, Reise,* p. 386.
—— ringens, *Bl. Schn.* p. 472.
—— piceus, *Poey.*
Melichthys ringens, *Bleek. Atl. Ichth.* v. pl. 220. f. 2; and *Ned. Tydschr. Dierk.* iii. 1865, p. 21.

Bleeker (who follows Swainson not only in adopting this as the type of a distinct genus, but in the classical error involved in the formation of the name) enumerates as one of the generic characters the smoothness of the caudal scales. In the Zanzibar specimen (13 inches long) there are eight series of caudal scales provided with spines directed forwards, and forming very distinct keels. Bloch also represents the keels in his figure.

Seychelles. *Indian Ocean and Archipelago. China. Pacific. Havanna.*

MONACANTHUS, *Cuv.*

465. **Monacanthus isogramma.** [306.]

Monacanthus isogramma, *Bleek. Nat. Tydschr. Ned. Ind.* xiii. p. 367 ; *Atl. Ichth.* v. pl. 222. f. 1 ; and *Ned. Tydschr. Dierk.* iii. 1865, p. 26.

Zanzibar. *Java.*

466. **Monacanthus pardalis.** [190.]

Monacanthus pardalis, *Rüpp. N. W. Fische*, p. 57, t. 15. f. 3.
Limonacanthus pardalis, *Bleek. Atl. Ichth.* v. pl. 230. f. 2 ; and *Ned. Tydschr. Dierk.* iii. 1865, p. 27.

Zanzibar. *Red Sea. Indian Ocean and Archipelago.*

467. **Monacanthus fronticinctus,** sp. n. Pl. XIX. fig. 2. [521, 665.]

D. 1 | 33. A. 30.

The height of the body, with the pelvic process stretched out, is about one-half of the total length without caudal. Snout moderately produced, its upper profile rather concave. Dorsal spine slender, as long as the distance between the eye and the mouth, slightly denticulated behind, the teeth being comparatively longer in smaller specimens than in adults. Free portion of tail a little higher than long, the sides having an elongated patch of setiform spines. In young specimens [521] the second dorsal ray is produced into a long filament as in *M. cirrhifer,* Schleg.

Colour dirty greenish brown, with several more or less distinct black longitudinal bands on the sides. *A black band across the forehead between the anterior margins of the orbits.* Fins dirty brown.

Length from 5 to 9 inches.

Zanzibar.

ALEUTERES, *Cuv.*

468. **Aleuteres monoceros.** [743.]

Balistes monoceros, *Osbeck.*
Aleuteres berardi, *Less. Zool. Voy. Coqu. Poiss.* pl. 7; *Richards. Ichth. Voy. Sulph.* p. 132, pl. 61. f. 1.
Alutera cinerea, *Schleg. Faun. Japon.* p. 292, pl. 131. f. 1.
Alutarius obliteratus, *Cant. Mal. Fish.* p. 353.
Aleuteres monoceros, *Bleek. Atl. Ichth.* v. pl. 226. f. 2 ; and *Ned. Tydschr. Dierk.* iii. 1865, p. 28.

Zanzibar. *East-Indian archipelago. China. Japan. Atlantic shores of America.*

469. **Aleuteres nasicornis.** [292.]

Alutera nasicornis, *Schleg. Faun. Japon.* p. 223, pl. 131. f. 2.
Pseudaleuteres nasicornis, *Bleek. Atl. Ichth.* v. pl. 221. f. 1, & 224. f. 2 ; and *Ned. Tydschr. Dierk.* iii. 1865, p. 28.

Zanzibar. *Mauritius. East-Indian archipelago. Japan.*

470. **Aleuteres scriptus.** [680.]

Balistes scriptus, *Osbeck, Voy.* i. p. 174.

—— lævis, *Bl. t.* 414.

Aleuteres lævis, *Richards. Ichth. Voy. Sulph.* p. 131, pl. 61. f. 3.

Alutarius lævis, *Cant. Mal. Fish.* p. 355.

Aleuteres scriptus, *Bleek. Atl. Ichth.* v. p. 227. f. 4; and *Ned. Tydschr. Dierk.* iii. 1865, p. 28.

Zanzibar. *East-Indian archipelago.*

Order LOPHOBRANCHII.

SOLENOSTOMA, *Lacép.*

471. **Solenostoma cyanopterum.** Plate XX. figs. 2, 3. [313.]

Solenostoma paradoxum, *Bleek. Nat. Tydschr. Ned. Ind.* iii. 1852, p. 308 (nec *Pall.*).

—— cyanopterus, *Bleek. Act. Soc. Scien. Indo-Neerl.* 1859, vi. p. 190.

D. 5 | 18. A. 16–17. P. 18. V. 7. C. 14.

Body much compressed; its greatest depth is below the origin of the dorsal, and is contained twice and a half or twice and two-thirds in the distance between the gill-opening and the root of the caudal. The body is narrower opposite the pectoral fin than opposite the origin of the dorsal; it is much attenuated above the ventral sac, and dilated again between the second dorsal and anal. The free portion of the tail is rather deeper than long. The length of the head equals the distance between the gill-opening and the end of the dorsal. Eye of moderate size, half the length of the postorbital portion of the head. Snout compressed into a thin lamella, its length being one-third of the total length without caudal; upper and lower edges of snout sharp, sides smooth, with the exception of a narrow stripe along its upper and lower edge, which is beset with fine vertical striæ. Edges of upper surface of crown raised, enclosing a groove, which is tapering in front. Operculum finely sculptured, with three raised lines radiating from the upper part of its base. The integuments of the body are divided into about twenty-five rings, the six or seven anterior of which are the broadest, extending to between the two dorsal fins; the remaining rings occupy the posterior half of the body and tail, and are much narrower. The divisions are formed externally by more or less conspicuous vertical ridges, intersected by three series of interrupted, short, longitudinal ones.

Pectoral fins very broad, composed of twenty-six short rays. First dorsal five-rayed, situated on the fifth dermal ring; the rays are very long, extending, when laid back, as far as the second dorsal fin. The second dorsal is composed of eighteen very small delicate rays, situated on a semicircular hump. Caudal very long, acutely rounded, composed of fourteen rays, the middle of which are nearly as long as the head. Anal with from sixteen to seventeen rays, very similar to the second dorsal. Ventrals seven-

T

rayed, very long, extending beyond the origin of the anal, coalesced with its inner side to the integuments of the body, both ventrals forming the egg-sac.

Ground-colour generally brown, minutely dotted with black and whitish. Dorsal fin with a large bluish-black ocellus between the first and second and second and third rays; the top of the fin of a beautiful purplish red, the remainder marbled like the body. Caudal with numerous small, oval, purple spots and dots.

In a second variety the markings are the same; but the ground-colour is rosy, and the markings are purplish brown.

Length from 4 to 5 inches.

Zanzibar. *East-Indian archipelago.*

Kaup (*Cat. Loph. Fish.* p. 2) states that in the males of *S. paradoxum* the egg-pouch is formed by the union of the inner edge of the ventrals to the skin of the belly, and that in the females the ventrals are free as in other fish. All the specimens from Zanzibar which have been examined have the ventrals attached to the skin of the belly, and *all* of them are *females*; so that if the first part of Kaup's remarks proves to be true, both sexes in this species carry eggs.

We may state that we have ascertained by dissection that specimens having eggs in the ventral pouch have at the same time ova in the ovaries scarcely less developed than those in the pouch.

PEGASUS, *L.*

472.　　　　　　　　**Pegasus draco.**　　　　　　　　[443.]

Pegasus draco, *L. Syst. Nat.* i. p. 418; *Bl.* t. 109. f. 1 & 2; *Gronov. Zoophyl.* no. 356, t. 12. f. 2 & 3;

Kaup, *Cat. Lophobr. Fish.* p. 5, pl. 1. f. 3 (not fig. 4 as stated in the list of plates).

Cataphractus draco, *Gronov. Syst.* ed. *Gray,* p. 144.

There are two specimens in this collection. The larger one is 2·5 inches long; and the projecting part of its snout, measured from the anterior rim of the orbit, is 0·3 inch; it is considerably dilated on its lower surface, forming an oval disk. The second specimen is 1·9 inch long; and the projecting part of the snout is 0·25 inch, consequently comparatively much longer than in the other; it is also much narrower, the projecting part being tetrahedral, with the sides equal and the edges strongly serrated. In other respects the specimens are alike.

Zanzibar. *East-Indian seas.*

HIPPOCAMPUS, *Cuv.*

473.　　　　　　　　**Hippocampus mannulus.**

Hippocampus mannulus, *Cant. Mal. Fish.* p. 388, pl. 11. f. 1.

This specimen appears to be *H. mannulus*, a species which apparently is distinguished by the frontal crest not terminating in a spinous projection in front.

Zanzibar. *Pinang.*

474.   **Hippocampus hystrix.**   [328.]

Hippocampus hystrix, *Kaup, Cat. Lophobr. Fish.* p. 17.

Zanzibar. *Japan.*

475.   **Hippocampus punctulatus.**

Hippocampus punctulatus, *Kaup, Cat. Lophobr. Fish.* p. 14, pl. 2. f. 1.

Zanzibar.

476.   **Hippocampus guttulatus.**   [137.]

Hippocampus guttulatus, *Cuv. Règne Anim.; Kaup, Cat. Lophobr. Fish.* p. 9, pl. 4. f. 3.

Aden. Zanzibar. *Mauritius. Gambia. Brazil.*

477.   **Hippocampus monckei.**   [324.]

Hippocampus monckei, *Bleek.?; Kaup, Cat. Lophobr. Fish.* p. 8.

Aden. Zanzibar. *Japan.*

478.   **Hippocampus subcoronatus,** sp. n.   Plate XX. fig. 4.   [609 A.]

This species is evidently closely allied to *H. coronatus*, Schleg.; but the occipital crest is much lower, and there are three spinous projections along each side of the base of the dorsal.

The length of the snout equals the distance from the hind margin of the orbit to the edge of the gill-covers, the interorbital crest projecting above its upper margin. Orbital and occipital spines well developed, simple. Occipital crest subtriangular, not quite as high as snout, and terminating in a subpentagonal knob. Body with eleven rings: the first, fourth, and seventh have the spinous prominences rather stronger than the others. Dorsal fin with eighteen rays; it is situated on the last two rings of the body and the two first of the tail. Caudal rings thirty-six in number, of which each alternate one has the spines more prominent.

Colour in a dried state.—Brown; head variegated with yellowish.—Length 4 inches.

Zanzibar. *Mozambique.*

SYNGNATHUS, *L.*

479.   **Syngnathus biaculeatus.**   [40.]

Syngnathus biaculeatus, *Bloch,* t. 121. f. 1; *Pet. Wiegm. Arch.* 1855, p. 277.

Gasterotokeus biaculeatus, *Kaup, Cat. Lophobr. Fish.* p. 19.

Aden. Zanzibar. *From the Red Sea, throughout the Indian Ocean, to China.*

480.   **Syngnathus fasciatus.**   [310.]

Syngnathus fasciatus, *Gray, Ill. Ind. Zool.; Pet. Wiegm. Arch.* 1855, p. 277.

Corythoichthys fasciatus, *Kaup, Cat. Lophobr. Fish.* p. 25.

Zanzibar. *Mozambique. Réunion. All Indian seas.*

481. Syngnathus zanzibarensis, sp. n. Plate XX. fig. 5. [312, 381.]

The length of the head in adult specimens is contained from ten to eleven times, and in immature specimens from eight to nine times in the total length. The length of the snout equals the distance from the anterior margin of the orbit to the extremity of the pectoral fins when laid backwards; it is scarcely compressed, and not much higher than broad. The diameter of the orbit is one-seventh of the length of the head, the space between the eyes is concave and less than the diameter of the orbit. Occiput somewhat elevated, without being raised into a ridge. Opercles swollen, without ridge, and nearly smooth. Trunk heptagonal, slender, twice and a third as long as the head, and, measured from extremity of snout to vent, half the length of the tail. The stomachic region is slightly thickened. There are twenty-one pairs of shields between the head and the dorsal fin, and as many between the throat and the vent. Tail tetrahedral, tapering, but not terminating in a point; the width of the upper surface is three-fourths of that of the lower one; the former is slightly, the latter distinctly concave. It has from fifty-nine to sixty-three rings, of which eighteen are occupied by the membranous egg-pouch in the males. The base of the dorsal is equal to the length of the snout, measured from the centre of the eye; it stands on three body- and three tail-rings, and consists of twenty-six rays, the length of which is less than the height of the tail beneath the last ray. Caudal very minute, more distinct in young than in adult examples.

Colour brown, marbled and punctulated with lighter and darker; young specimens have about thirteen broad brown cross bands, at equal distances, occupying the whole of the lower surface of the trunk.—Length from 4 to 13 inches.

Zanzibar.

482. Syngnathus mossambicus. [311.]

Syngnathus mossambicus, *Pet. Wiegm. Arch.* 1855, p. 277.

The description of this fish given by Peters agrees perfectly with our specimens, except that the head would appear to be relatively about twice as long; he states that in a specimen 140 millimetres long, the head was 7 millimetres only; we suppose that this is a misprint for 17 millimetres.

Zanzibar. *Mozambique.*

Order CHONDROPTERYGII.

STEGOSTOMA, *Müll. & Henle.*

483. Stegostoma fasciatum. [132, 721, 513.]

Seba, iii. p. 105, t. 34. no. 1.
Squalus fasciatus, *Bl.* t. 113.

Squalus tigrinus, *L. Gm.* i. p. 1493. no. 19.
— longicaudus, *L. Gm.* i. p. 1496. no. 24.
Russell, pl. 18 (young).
Scyllium heptagonum, *Rüpp. N. W. Fische*, pp. 61, 74, pl. 17. f. 1.
Stegostoma carinatum, *Blyth, Journ. Asiat. Soc. Ben.* 1847, p. 725, pl. 25. f. 1 & 1 a.
—— fasciatum, *Müll. & Henle, Plag.* p. 24; *Cant. Mal. Fish.* p. 396; *Duméril, Hist. des Poiss.* i.
      p. 337.

There are three specimens in this collection:—one [132], measuring 10 inches, corresponding to Russell's plate; another [721] corresponding to Cantor's "older variety," length 2 feet; and the third [513], of the same size as the second, which appears to be an undescribed variety.

It is identical in structure with the ordinary specimens; but the coloration is very different. Taking the light yellow of the under portions as the ground-colour, the upper and lateral parts of the head and body are densely ornamented with undulating brown stripes, darker on the edges than in the middle. On the tail these vermiculated stripes assume the appearance of more or less regular transverse bands, from forty to fifty in number, counting from the second dorsal. The sides of the belly are yellowish, minutely dotted with brown; the lower portions are immaculate.

Aden. Zanzibar. *Madagascar. Indian Ocean and Archipelago.*

GINGLYMOSTOMA, *Müll. & Henle.*

484.   Ginglymostoma brevicaudatum, sp. n. Plate XXI.   [710.]

Head very flat, depressed. Snout short, rounded, obtuse. Caudal comparatively short, its length being contained four times and a half in the total.

The nasal margin of the upper lip is provided in its middle with a very short cirrus. The isthmus between the postlabial folds of the lower jaw is about as wide as each of these folds. Eye very small. Spiracle minute, situated close behind the eye. The angles of all the fins are rounded. The dorsal and anal fins are equal in size; the origin of the first dorsal is a little behind the middle of the base of the ventral: the interspace between the two dorsals is scarcely longer than the base of the first; the second dorsal commences somewhat before the anal, and terminates a little before the posterior end of that fin. Posterior margin of caudal excised; its lower lobe is obtusely rounded. Pectorals subquadrangular, with the posterior margin very much rounded. Scales distinctly larger than in *G. cirratum*; those on the head are smooth; those on the back have from one to three keels, which become spinous on the scales of the side.

Upper parts dark brown, with minute black dots; the lower surface brownish white. Total length 25 inches. Distance of vent from extremity of snout 12 inches. Zanzibar.

CARCHARIAS, *Cuv.*

485. **Carcharias melanopterus.** [105, 130.]

? Squalus carcharias minor, *Forsk.* p. 30.
Carcharias melanopterus, *Quoy & Gaim. Voy. Uran. Poiss.* pl. 43. f. 1, 2; *Benn. in Life of Raffles,*
  p. 693; *Rüpp. N. W. Fische,* p. 63.
Carcharias (Prionodon) melanopterus, *Müll. & Henle, Plag.* 43; *Richards. Ichth. China,* p. 194;
  Duméril, Hist. des Poiss. i. p. 365.
Prionace melanopterus, *Cant. Mal. Fish.* p. 400.

The distinguishing character of this species is the deep-black tips to all the fins.

Aden. *Red Sea. East-Indian archipelago. Australia. Timor.*

ZYGÆNA, *Cuv.*

486. **Zygæna malleus.**

Squalus zygæna, *L. Syst. Nat.* 399.
Russell, pl. 12.
Zygæna malleus, *Shaw, Nat. Misc.* pl. 267; *Val. Mém. du Mus.* 1822, t. ix. p. 223, pl. 11. f. 1, a & b;
  Schleg. Faun. Japon. p. 306, pl. 138.
Sphyrna zygæna, *Müll. & Henle, Plag.* p. 51; *Bonap. Faun. Ital. Pisc.*
Cestracion zygæna, *Duméril, Hist. des Poiss.* i. p. 382.

Aden. *Zanzibar.* Seychelles. *Mediterranean. Indian Ocean. Atlantic.*

487. **Zygæna tudes.** [134.]

Zygæna tudes, *Cuv. Règne Anim.; Val. Mém. du Mus.* ix. p. 225, pl. 12. fig. 1, a & b.
Cestracion tudes, *Duméril, Hist. des Poiss.* i. p. 384.

Aden. *Mediterranean. Atlantic.*

PRISTIS, *Lath.*

488. **Pristis peroteti.**

Pristis peroteti, *Müll. & Henle,* p. 108; *Duméril, Hist. des Poiss.* i. p. 474.

Zanzibar. West coast of Africa.

RHINOBATUS, *Müll. & Henle.*

489. **Rhinobatus schlegelii.** [459.]

Rhinobatus schlegelii, *Müll. & Henle, Plag.* p. 123, t. 42; *Richards. Ichth. China,* p. 195; *Schleg.*
  Faun. Japon. Poiss. p. 307; *Duméril, Hist. des Poiss.* i. p. 497.

Zanzibar. Japan.

TRYGON, *Müll. & Henle.*

490.   **Trygon uarnak.**   [713.]

Raja uarnak, *Forsk.* p. 18. no 16 b.
Pastinachus uarnak, *Rüpp. N. W. Fische,* p. 69, t. 19. f. 2.
Trygon uarnak, *Müll. & Henle, Plag.* p. 158; *Cant. Mal. Fish.* p. 513.
Trygon (Himantura) uarnak, *Duméril, Hist. des Poiss.* i. p. 585.
Zanzibar. *Red Sea. Seychelles. Indian and China seas. East-Indian archipelago. Cape of Good Hope.*

491.   **Trygon pastinaca.**   [552.]

Raya pastinaca, *L. Syst. Nat.* p. 396.
Trygon pastinaca, *Cuv. Règne Anim.*

The two specimens in this collection from Zanzibar have large blue ocelli with lighter centres.
Zanzibar. *Cape of Good Hope. Mediterranean. British Channel.*

TÆNIURA, *Müll. & Henle.*

492.   **Tæniura lymma.**   [42.]

Raja lymma, *Forsk.* p. 17. no. 15.
Trigon lymma, *Cuv. Règne Anim.;* *Rüpp. Atlas,* p. 51, t. 13. f. 1, and *N. W. Fische,* p. 69, t. 19. f. 4 (teeth); *Cant. Mal. Fish.* p. 520.
—— ornata, *Gray, Ill. Ind. Zool.*
—— halgani, *Less. Voy. Coqu.* p. 100, pl. 2.
Tæniura lymma, *Müll. & Henle, Plag.* p. 171; *Duméril, Hist. des Poiss.* i. p. 619.
Aden. Zanzibar. Mozambique. *Red Sea. East-Indian archipelago. New Ireland.*

TORPEDO, *C. Dum.*

493.   **Torpedo fuscomaculata.**   [379. 380.]

Torpedo fuscomaculata, *Peters, Wiegm. Arch.* 1855, p. 278.
The larger of the two specimens in the collection, measuring 10 inches, is spotted with dark brown, much in the way described by Peters; but the smaller specimen, measuring 7 inches, is perfectly immaculate, having, however, the white edges to the fins.
Mozambique.

CEPHALOPTERA, *C. Dum.*

494.   **Cephaloptera kuhlii.**   [554.]

Cephaloptera kuhlii, *Müll. & Henle,* p. 185, pl. 59. f. 1; *Duméril, Hist. des Poiss.* p. 654.
Zanzibar. *Indian Ocean. Amboyna.*

ADDENDA.

Page 17, after **Mesoprion bohar**, add :—

34*. **Mesoprion argentimaculatus.** [094.]

Sciæna argentimaculata, *Forsk.* p. 50.
Diacope argentimaculata, *Cuv. & Val.* ii. p. 432 ; *Rüpp. Atl. Fische*, p. 71, t. 10. f. 1.
Mesoprion argentimaculatus, *Günth. Fish.* i. p. 192.

Seychelles. *Red Sea.*

Page 31, after **Heterognathodon petersii**, add :—

87*. **Heterognathodon flaviventris.**

Heterognathodon flaviventris, *Steindachner, Verhandl. zool.-bot. Ges. Wien*, 1866, p.778, pl.13. fig. 6*.

$$D. \frac{10}{6}. \quad A. \frac{3}{7}. \quad L. \text{ lat. } 46. \quad L. \text{ transv. } \frac{3\frac{1}{2}}{0-10}.$$

Length of the head contained twice and three-fourths in the total without caudal, and equal to the height of the body. A yellow band on each side of the belly, between the mandible and the caudal. A greenish-yellow longitudinal band above the base of the dorsal.

Diameter of the eye contained thrice and three-fifths in the length of the head. Width of the forehead two-elevenths, and the length of the snout two-sevenths of the length of the head. Maxillary extending to below the middle of orbit. There are three vertical series of scales between the eye and præopercular angle. Præoperculum finely serrated behind and slightly concave. Last dorsal spine longest, equal to the distance of the extremity of the operculum from the posterior margin of orbit. The last dorsal ray extending beyond the base of the caudal. Third anal spine longest. Pectoral seven-eighths of the length of the head, extending to the first soft ray of the anal. Caudal deeply notched, with the upper lobe four-fifths of the length of the head. Scales minutely serrated. (*Steindachner.*)

Zanzibar.—Length 5½ inches.

Page 45, after **Lethrinus longirostris**, add :—

137*. **Lethrinus genivittatus.**

Lethrinus genivittatus, *Cuv. & Val.* vi. p.306, pl.159 ; *Steindachner*, p. 478.

Zanzibar.

* The volumes containing these descriptions has not yet been published ; the references are taken from a separate copy sent by the author.

Page 45, after **Lethrinus nebulosus**, add :—

140°. **Lethrinus striatus ?**

Lethrinus striatus, *Steindachner, Verhandl. zool.-bot. Ges. Wien*, 1866, p. 479, pl. 5. fig. 3.

D. $\frac{10}{9}$. A. $\frac{3}{8}$. L. lat. 45. L. transv. 5/14.

Closely allied to *Lethrinus nebulosus*, but perhaps essentially different by the larger size of the scales and of the head and coloration.

Length of the head contained thrice and three-fourths in the total; height of the body a little more than thrice. Diameter of the eye nearly one-fourth of the head; length of the snout not quite one-half. Width of forehead nearly equal to the diameter of the eye. Upper molar teeth somewhat larger than the lower. Dorsal spines very strong, the fifth being the longest and contained about once and a third in the length of the middle rays, or twice and a third in that of the head. The third broad anal spine considerably longer than the two anterior ones, which also are very strong. Pectorals and ventrals extending to the base of the first anal spine. Caudal moderately emarginate, the longest rays being equal to the distance of the extremity of the snout from the posterior margin of the orbit.

A blackish-brown band across the middle of the forehead; a second, arched; one from the middle of the front margin of the orbit to the middle of the snout. Bluish-violet bands along the series of scales above the lateral line. (*Steindachner.*)

Zanzibar.

Page 53.

175. **Histiophorus brevirostris.**

Since the sheet containing the description of this species was printed, the typical specimen, or rather the remains of it, consisting of the complete head, fins, and portion of the skin, have been received at the British Museum. This has induced us to reexamine it, and especially to compare it with *Tetrapterus lessonii*, described by Canestrini in 'Archivio per la Zoologia,' 1861, vol. i. p. 259, pl. 7, which fish was found in the Mediterranean.

There cannot be a doubt that the two fish are extremely closely allied—so much so that, had we not already given the Zanzibar specimen a new name, we should have hesitated to separate the two. Besides the difference of habitat, however, there are others which may tend to prove their real distinctness. The head of the Mediterranean species is comparatively longer, the anterior part of the dorsal much higher, and the pectoral longer. There are also differences in the number of fin-rays. The Zanzibar specimen has D. 38 | 7. A. 12 | 7; that from the Mediterranean D. 44? | 7. A. 17?* | 6.

* Is this number not a misprint? The figure indicates 12 only.

Page 70, after **Gobius giuris**, add:—

236*.   **Gobius obscurus.**

Gobius obscurus, *Peters, Wiegm. Arch.* 1855, p. 250; *Steindachner, Verhandl. zool.-bot. Ges. Wien,* 1866, p. 780, pl. 18. f. 6.

Zanzibar. *Mozambique.*

Page 71, to description of **Gobius caninus**, add:—

The African variety of this species appears to have been described by Dr. Steindachner (Verhandl. zool.-bot. Ges. Wien, 1866, p. 781, pl. 18. fig. 17) as a distinct species under the name of *G. petersii.* A renewed examination of the specimens in the British Museum has not induced us to change our views as already expressed.

Page 82, after **Pomacentrus punctatus**, add:—

283*.   **Pomacentrus trichourus**, sp. n.   Plate XVII. fig. 5.   [484.]

$$D. \frac{14\text{--}15}{13\text{--}14}. \quad A. \frac{2}{14\text{--}15}. \quad L.\ lat.\ 25. \quad L.\ transv.\ 3/10.$$

The height of the body is half of the total length (the caudal not included). The length of the head is contained thrice and a half in the same. Interorbital space scarcely convex, equal to the diameter of the orbit. Præorbital irregularly serrated, half the width of the orbit. Præoperculum coarsely denticulated. The dorsal spines increasing slightly in length posteriorly, the last being five-eighths of the length of the head. Caudal slightly emarginate. The second anal spine strong, and equal in length to one of the middle spines of the dorsal. The first ventral ray produced into a short filament. The free portion of the tail somewhat higher than long.

Colour in life bluish grey, each scale with two bright blue spots, after death uniform brownish. Dorsal and anal with brown margins. Caudal bright yellow, with a broad greyish margin and a blackish base. A bright blue spot on the superior angle of operculum, becoming black in a preserved state. Base of pectorals black.—Length 1½ inches.

Zanzibar.

INDEX.

THE END.

PRINTED BY TAYLOR AND FRANCIS,
RED LION COURT, FLEET STREET.

x

www.ingramcontent.com/pod-product-compliance
Lightning Source LLC
Chambersburg PA
CBHW021808190326
41518CB00007B/506